The Gifford Lectures 1972/73

The Development of Mind

A. J. P. Kenny, *Balliol College, Oxford*
H. C. Longuet-Higgins FRS, *University of Edinburgh*
J. R. Lucas, *Merton College, Oxford*
C. H. Waddington FRS, *University of Edinburgh*

Edinburgh
at the University Press

© 1973
EDINBURGH UNIVERSITY PRESS
22 George Square, Edinburgh
ISBN 0 85224 263 8
North America
Aldine Publishing Company
529 South Wabash Avenue, Chicago
Printed in Great Britain by
R & R Clark Ltd, Edinburgh

Contents

ONE
H. C. LONGUET-HIGGINS
The Frontiers of Psychology 1

TWO
J. R. LUCAS
Explanations of Mind 16

THREE
C. H. WADDINGTON
The Development of Mind 30

FOUR
A. J. P. KENNY
The Origin of the Soul 46

FIVE
LUCAS, LONGUET-HIGGINS, WADDINGTON, KENNY
Open Discussion 61

SIX
C. H. WADDINGTON
The Evolution of Mind 74

SEVEN
A. J. P. KENNY
The Origin of Language 91

EIGHT
H. C. LONGUET-HIGGINS
Possible Minds 108

NINE
J. R. LUCAS
The Genesis of Mind 123

TEN
KENNY, LONGUET-HIGGINS, LUCAS, WADDINGTON
Questions and Answers 137

Preface

In May 1969 the Principal of Edinburgh University wrote to J. R. Lucas in these terms:

Dear Mr Lucas,

You will, I am sure, know what our Gifford Lectures are. In the past we have nearly always invited one lecturer to give a double series, that is to say, 10 lectures one year and 10 lectures the next year, and these are normally printed as a book.

The Gifford Committee, however, is disposed for the period 1971/72 and 1972/73 to try something different. Our idea, which is by no means fully formulated, is to persuade, if we can, four people to combine, giving perhaps five lectures each, with, we would hope, some joint seminars or disputations, the whole thing to be a developing concept. The general theme would be in the very broadest sense the development of mind, and we hope to start with a physical scientist, to move on with an animal behaviourist, or a psychologist, then some sort of philosopher, and finally a theologian.

We now have in the University as a Royal Society Professor someone who is, I believe, a friend of yours, namely Christopher Longuet-Higgins, who is much concerned nowadays with the physical basis of mind, and he has agreed to take part. We have just had a further meeting of the Committee with him, as a result of which we felt unanimously that we should like to ask you to be the theological participant. Although we had a number of ideas of biologists, psychologists, and philosophers of one sort or another, we didn't feel quite sure who would fit in best, and in the end we agreed that I should approach you to see whether you were in any way interested, and, if you were, further to invite you to get together with Christopher Longuet-Higgins and decide what other people you might like to rope in. . . .

Yours sincerely,
Michael Swann

The invitation was accepted, and before long the team was completed by the recruitment of A. J. P. Kenny, representing philosophy, and C. H. Waddington, representing biology. The individual participants gave notice that they might step outside their allotted roles, as indeed they did when the time came. It was agreed to entitle the two-year series *The Phenomenon of Mind*; the title of the 1971/72 lectures would be *The Nature of Mind* and of the 1972/73 lectures *The Development of Mind*.

The lectures comprising *The Nature of Mind* were published in 1972; the second series of lectures was duly given, and the present volume is a record of them. In writing them up we have been a little kinder to ourselves than with the first series, and have attempted to transform our more rambling remarks into acceptable prose. As in the previous year, most of the lectures began with a formal presentation by one of the lecturers, followed by a discussion introduced by another who had seen the detailed text of the prepared talk. But in the fifth lecture we attempted to answer questions submitted in advance by members of the audience, and in the last lecture we took it in turn to answer questions put to us by our three colleagues.

As before we would like to express our sincere thanks to Sir Michael Swann, and to our other colleagues who took the Chair. Our special thanks are due to Miss Constance Masterton, the secretary of the Gifford Committee, for her tact and efficiency in making all the practical arrangements.

A. J. P. Kenny
H. C. Longuet-Higgins
J. R. Lucas
C. H. Waddington.

First Lecture. The Frontiers of Psychology

In last year's Gifford Lectures my colleagues and I discussed *The Nature of Mind*. How much light we cast on the matter we must leave you to judge. None of us, as Waddington remarked in the last lecture but one, attempted to define 'mind'; but I think we all dropped pretty heavy hints as to what we were talking about. John Lucas laid great stress on his rationality and moral autonomy. Tony Kenny set much store by his ability to speak and understand a natural language. Wad aligned himself with all those creatures which select and modify their own goals. And I claimed that, even if computers can't think, at least you and I can compute.

This year our subject is *The Development of Mind*. In case, by some oversight, we should fail to define 'development', let me say at once how I shall be using the word. I want to use it in its biological sense − or, rather, in its two biological senses. The first sense occurs in the name developmental psychology, which is the study of mental development in the individual from conception to senility. I deliberately say 'mental development' rather than 'behavioural development', because like Chomsky I value the distinction between subject-matter and evidence. A baby's behaviour may be a very good clue to its mental processes, but this does not imply that the object of our enquiry is the behaviour itself. In any case other kinds of evidence are also available to us; such as the baby's heart rate which goes up suddenly, so Tom Bower credibly asserts, when the baby is surprised. And if surprise is a form of behaviour rather than a mental state, then I am not a native speaker of English.

The other biological sense of the word 'development' is the evolutionary sense. Two billion years ago, we believe, there was a primaeval soup in which the first living things assembled themselves. Whatever one's views about the nature of mind, one can hardly deny that there is a good deal more mental activity going on now than there was then. What has caused this remarkable development?

The two questions − why do we develop minds as we grow up, and why has evolution produced minds at all − will be my subject. I shall suggest that the former question is a perfectly sensible one, but that we just don't happen to have much of an answer to it. We know, thanks to Piaget and his successors, quite a lot about *how* a child's mind develops, but we have little or no idea *why* it develops further than that of a chimpanzee, say. The other question, *why* did evolution

1

produce beings with minds, I believe to be unanswerable, for reasons which I shall explain. But I see no reason to doubt that we shall discover *how* it happened; and furthermore, that when we know the evolutionary history of our species, we shall be in a position to understand *why* our minds develop along the lines that the psychologists are beginning to map out.

I must explain what I mean by 'how' and 'why'. I remember being told by my elders and betters that science could answer 'how' questions but not 'why' questions, and I believed this for many years. The suggestion was that science offers us general laws about the *way* in which things happen, but gives no *reasons* why these laws should hold rather than others which we might think up. In a sense this is true, but it misses one very important point: that the laws of science are not logically independent assertions about how the world behaves, but have to fit together as far as possible. The chief difference, between natural science on the one hand and unbridled speculation on the other, is that scientific assertions must harmonise, not only with experience, but with one another. If an otherwise plausible hypothesis in biology clashes with one in physics, then one or other must go by the board. Dissension can be tolerated for a while, but not indefinitely.

The fact that science is a coherent body of ideas, and not simply a disconnected set of observations, makes it possible for science to answer not only 'how' questions, but also certain sorts of 'why' questions. The answer to a 'why' question will require an appeal to laws more general than those requiring explanation, and there will come a point at which the laws being appealed to cannot be explained scientifically. But a very few primary laws may suffice to account for a very wide range of events and phenomena, and the most remarkable thing about the physical world is the fewness of the principles on which it seems to operate. I say 'the physical world' advisedly, because it is not at all obvious that the same must be true of the living world and the world of human affairs. So we had better consider the relation between these worlds, or rather, between the ideas which we use for thinking about them.

Last year I talked at some length about 'reductionism', and criticised the suggestion that psychology was really physiology, which was really chemistry, which was really physics, so that psychology was really physics. I now realise that this simple message was open to misunderstanding. In his book, *Beyond Reductionism,* Koestler seems to suggest that we can expect little help from physiology in our attempts to understand how we think or perceive; and this I would wish to deny. But of course the relation between a pair of sciences, such as psychology and physiology, or chemistry and physics, which obviously border one another, needs very careful definition. What I was trying to say last year was that the concepts of the 'higher' science — adopting the convention that psychology is 'higher' than physiology — cannot be arrived at by analysing those of the lower science. What I would now add is that although psychological questions must be *posed* in psychological terms, questions about the mechanism of thought or perception can only be *answered* in physiological terms. In other words, if the questions which arise in any science cannot be answered within that science, they can only be answered by a lower science, not by a higher one. It would, for example, seem distinctly strange to offer a psychological explanation for the propagation of impulses along nerve fibres!

With these warnings in mind, let us see what modern psychology can tell us about the mental development of the human child. Here I tread with the utmost delicacy, because Edinburgh is a centre of excellence in the study of babies and small children. The first problem is to find a way of describing a child's mental faculties at any stage in its development: what concepts it has at its disposal, what perceptual tasks it is able to perform, and how it sets about solving practical problems such as arranging wooden sticks in order of length. The greatest challenge to scientific description is the astonishing feat of first language acquisition; and once the child, at three or four years old, is able to talk to the psychologist the range of possible studies is enormously increased, and so is the quality of the available evidence. I will not attempt to review our present knowledge of human mental development, particularly as Waddington will be talking later about the work of Jean Piaget. I will only say that in Piaget's view all children go through very much the same stages, in the same order, just as we all go through the same stages of physical development, in the same order. The difficulty with this assertion is not in believing it, which is very easy, but in substantiating it, which is very difficult: how is one to define a 'stage', so that one can recognise without doubt when two children are at the same stage? Piaget's special achievement, as I understand it, has been so characterise these stages, or 'structures' as they are called by the Geneva school, in terms which are relatively unmistakable to the trained observer.

In passing, I must admit to a slight unease about the use of the word 'structure' in this connection. The word suggests something entirely passive, like a bit of scaffolding. Surely psychological development consists of the progressive mastery of different skills, and of the concepts required for them. For this reason I am attracted by Seymour Papert's interpretation of Piaget's 'structures' as programs* or routines which arise in

* The word 'program' is a technical computing term, not the
  American spelling of 'programme'.

3

the child's mind and are to some extent, but not fully, open to introspection and conscious modification. Papert, on the basis of his own observations, believes that one can teach children to 'debug' their own thinking programs by teaching them, at a very early age, to write and run simple programs on computers. This might turn out to be the best use to which computers can be put.

For those who are not familiar with the jargon of computing I should explain that a 'bug' is simply a programming error. The most insidious bugs are those undetected errors which cause a program to fail intermittently. One has to be continually debugging one's mental programs. When we went over to decimal currency, most people managed to re-program their monetary thinking without too much trouble, except that a lot of people had difficulty in persuading themselves that the florin was worth only ten new pence, not twenty. In my case I am pretty sure that this was because of the visual resemblance between the symbol for two shillings and the symbol for twenty; as soon as I spotted this bug I stopped making that particular mistake. But the sixty-four-dollar question — to think in American currency for the moment — is this: what causes Piaget's structures or Papert's programs to arise in the child's mind in the first place?

The answer to this question might be superficial, or it might be beyond the reach of scientific investigation; I believe it is neither. It might be superficial if all our mental skills were directly imparted to us by our parents and teachers, as some of our skills undoubtedly are: for example, the kind that an apprentice learns from a master craftsman. But there is overwhelming evidence that our most primitive skills, such as our ability to interpret visual impressions in spatial terms, and our ability to learn the grammar of our native language, are inborn. Psycholinguists have made much of the apparent uniqueness of the human being in his ability to discover the grammar of his parents' language. Whether we are indeed unique in this respect doesn't seem to me to matter very much; I would be more than happy to think that a chimpanzee such as Sarah or Washoe could learn a language of the kind which human beings use. It does seem extremely unlikely, though, that we could ever teach cats or rabbits to read, or even to communicate with us in sign language; one's pessimism is based on the feeling that these animals lack some absolutely basic mental capacity without which they could never even get to Square One. It is this mental capacity, that Chomsky describes as our knowledge of universal grammar, which we are fortunate enough to bring into the world with us.

The trend in developmental psychology at the moment seems to be away from the idea that the human being is little more than the product of a lifelong schedule of operant conditioning, and towards the view that his mind, just like his

4

body, develops largely from the inside, as it were. It would be most surprising if this were not so, because the vehicle of our thoughts, namely the central nervous system, continues to grow and ramify, though more and more slowly, during the first few years of our lives. The distinguished psycholinguist Jacques Mehler, alive to the acute difficulties facing any 'instructive' theory of mental development, has gone so far as to suggest that the new-born child knows all the answers, as it were, and merely forgets those which turn out to be irrelevant to his circumstances. Perhaps it is a sign of the times that such an idea should be advanced in the name of science; it was actually expressed by William Wordsworth nearly two centuries ago.

The conclusion to be drawn from these remarks is that the development of the child's mind seems to be entirely in line with what we know about the development of other biological functions in a growing organism. All animals, as they grow from birth to maturity, develop not only in stature but in wisdom: the particular kind of wisdom they need for the particular styles of life they lead. Though a philosopher might run into dualistic difficulties about the relation between the animal's physical growth and its mental development, the zoologist suffers from no such hang-ups. He takes it as a matter of course that every biological function — and thinking is no exception — must be performed by some organ, or organs, and that until these are fully developed the function must remain immature or unrealised. But there is something else which the zoologist knows, and which is not at all obvious to a mere pet-fancier. He knows that the growth of an animal — and indeed of any organism — is controlled and directed by a program, immensely longer and more informative than any program ever fed into a computer. The longest man-made program I know of is about as long, on paper, as an ordinary detective story. The program for constructing a human being in the womb runs to about the length of the *Encyclopeadia Britannica*. The program for making me is written in every cell of my body; if you looked you could just see, with a powerful microscope, the reels of tape on which it is written, though the individual letters are far too small. So if we are ever to understand in full detail why our brains, and hence our minds, develop as they do, we shall have to understand the instructions which are written on these reels of tape: the chromosomal DNA.

Put in this way, the problem of understanding how our bodies and minds are formed looks totally insoluble: even if we could read the billion letters of a human blueprint, how could we ever make sense of them? And so it would be, were it not for the other things that the biologist has to tell us, not only about the way in which organisms grow, but also about their evolution.

5

Everyone knows the main outlines of the theory of evolution. The problem, very roughly speaking, is to understand why there are men and dogs about, but not dodos or dinosaurs. The reason is simple: the dodos died out, and the men haven't yet, presumably because intelligence has so far proved a more valuable asset to human beings than elegance to the dodos. The problem reduces to that of explaining how a species can, as it were, conduct experiments in survival. Nowadays we know that the range of experiments which a species can try out is essentially determined by its genetic resources, enshrined in the chromosomal DNA of all individuals of the species. The DNA is pretty faithfully transmitted from one generation to the next, but now and again there are errors of transcription, called mutations, which may result in the appearance of a new breed. So the history of any existing species, such as our own, is to be thought of as a random walk through a space of possible forms, with the special property that every form along the route had to be fully viable.

Now this constraint, that all one's ancestors must have been fit enough to survive to maturity, has an analogue in developmental biology; namely that at any stage before an individual reaches maturity he must be in good enough working order, biologically speaking, to reach the next stage. And this kind of constraint is not unique to biology; it is thoroughly familiar to the civil engineer who has to make sure not only that his bridges will not blow down when built, but that they will not collapse during construction. What I am leading up to is the thought that it may be an extremely tricky problem — fortunately Nature's problem, not ours — to put together a highly complex organism; and perhaps the only way of doing so will be to follow pretty closely the ancestral walk through the space of possible forms, starting with the most primitive and ending with the most mature. This is the thought which enables us to understand dimly why ontogeny, the development of the individual, roughly recapitulates phylogeny, the evolution of the species.

The structure of my main argument may now be apparent to you. If it is really true that ontogeny must largely recapitulate phylogeny, then we do not have to wait until the molecular biologist has read our chromosomes before we can begin to understand why our minds develop as they do. We may, instead, be able to gain some enlightenment from studying our evolutionary ancestors and cousins. The missing links recently found by Robert Leakey in East Africa, and the partial success of the Premacks in teaching Sarah to read and write, hold up the mirror to our nature much better than any molecular biological study possibly could; though without molecular biology we would be unsure of the validity of such comparative studies, because we would not properly understand the evolutionary process.

6

Let me expand briefly on the way in which an evolutionary view might affect our thinking about mental development in human beings. One immediate consequence might be that we took a distinctly less academic view of such concepts as intelligence, rationality, morality and so forth. Take, for example, intelligence. If the human mind is the product of evolution by natural selection, then the quality for which Nature has selected us is probably not the capacity for abstract thought but the capacity to use our heads in tricky situations. On this view, the poet, the philosopher and the scientist enjoy faculties which have been earned for them by the hunter, the craftsman and the man of action. A romantic and a salutary thought. Or, if we are thinking about morality, we may surmise that the survival of our species has in the past depended upon its members helping and protecting one another, rather than merely telling one another not to do this or that. But here I am quite out of my depth, knowing only that the moral sense is a very delicate plant which is all too easily blighted by harsh conditions in early childhood.

The only trouble with the evolutionary approach is that *homo sapiens* now seems to be evolving in a different way from any other species. Claims for the uniqueness of man might have been open to question two million years ago; but whether it is to our credit or not, we seem to have set a genuine precedent in inventing the idea of a civilisation, or cultural tradition. The existence of civilisation rests crucially upon our ability to record our thoughts in writing and, more recently, in other media as well. We seem to be no longer dependent upon the vicissitudes of genetic variation, but to have replaced them by the vagaries of human politics. It seems that we are on our own now; having taken over from Nature we must accept the responsibility for moulding our biological future.

In introducing my main thesis I remarked that we already know, through the work of the developmental psychologists, quite a lot about *how* the child's mind develops but that we should not succeed in understanding *why* it develops as it does until we had acquired a better evolutionary perspective. I hope my reasons for this assertion are now rather clearer. The development of our brains, like that of our bodies, is very largely controlled by the programs embodied in our chromosomal DNA; and these programs record the biological secrets which enabled our ancestors to escape extinction. So if we are concerned to understand the most striking features of our own mental development, such as our ability to master our first language, we should look not only at the way our brains develop but also at our recent evolutionary history. At present we can only observe that our brains continue to develop both in size and in interconnectedness as we advance from infancy to maturity. But how well such a pattern of physical development, leaving adequate scope for the impressions of early childhood, would suit the needs of a species whose biological prosperity depended upon the handing on of a cultural tradition!

The question *why* evolution has produced creatures capable of talking English, building cities and visiting the Moon, is an altogether more forbidding one. All the evidence points to the view that nothing can create new chromosomal DNA except completely random events; this is the Central Dogma of molecular biology. On the other hand, an initially random variation will tend to be perpetuated if it suits the established life cycle of an animal, which is why species tend to get better and better at what they are already good at. So once a species has found a particular niche, it will tend to exploit that niche to the utmost. Unfortunately – or perhaps fortunately – this principle seems to be of little or no predictive power. Organisms have a way of discovering niches and exploiting them in ways which no-one would have thought of, just as, in our own era, history has a way of taking totally unexpected turns which only afterwards do we recognise – if we are good Marxists – for their historical inevitability. All one could have confidently said, in surveying the primaeval soup, was that evolution would almost certainly produce something interesting. We like to think that human beings are specially interesting.

Whether this is an acceptable proposition in natural theology or not, I must leave my colleagues to discuss. At least it is very nice to know that no scientific obstacle to our existence would be posed by the lack of a Creator with our interests specially in mind. In other words, there is much to be said for having a theory of the origin of mind which does not involve us in an infinite regress. In this respect the origin of mind is a similar problem to the origin of life; it would be commonly regarded as a confession of intellectual defeat to conclude that life on earth must have come from somewhere else.

If I stopped at precisely this point you might suspect me of exceptional insensitivity to the mysteries of our existence. So I will say one more thing before I close. There are different ways of talking about Man and Nature; some of them emphasise Man and some Nature. Traditionally, the scientist treats of Nature and of Nature's indifference to Man; and in much older tradition, the poet sings of Man and of his dependence upon Nature. Need we shut our ears to either?

Discussion
LUCAS
Let me start with points on which I agree with Christopher. The point I found most interesting was his bridge-building analogy which explains the recapitulation, in the development of the individual, of the development of the species. I think that the account he gave was a valuable and illuminating one,

and one that has a certain moral for the philosopher; in particular, although this is not an entirely new point, that we should not think of the mind simply, as John Locke thought, as a blank sheet of paper waiting for the impress of outside experience, but rather as an internally programmed — although, as I would say, not entirely programmed — system which generates a great deal of our intellectual activity. Particularly to those studying philosophy, who are brought up on the English Empiricists, I stress the point how wrong and how unempirical the Empiricists were. We don't sit and wait for things to be impressed on our consciousness: we try out things and see what happens. That is to say, we are always putting Nature to the question; we set the questions, but Nature gives the answers. In this respect, I am disagreeing with John Locke, and I think it is an important point to bring out.

There is another thing Christopher said, which makes a point Locke often made, namely, the relative unimportance of academic activities. Locke, putting it in a theological mode which Christopher would not adopt, said that God gave us sufficient understanding for the practical affairs of life, but not so as to fit us to speculate on matters with which we had no proper concern. Perhaps that goes a bit too far. What is, however, of great importance is that when we come to try and understand the mind, we too easily take an excessively academic attitude to it; and then with Kant we find soon that it is necessary to abolish knowledge in order to make room for faith. The proper answer to this is to remember that Kant offers us a critique of only the *pure* reason, and that Christopher is asking us to do what a study of evolution is forcing him to do; to consider the mind first and foremost as a practical matter. It is a matter of making choices, when you are hunting, shooting, or carrying on some line of business; it is a matter of choosing aright. And this is a second point which I feel we could very well learn from the biologists.

Now for some disagreements. When Christopher was giving his account of the evolution of mind (p. 8), he gave us the Central Dogma: nothing can create new chromosomal DNA, except completely random events. A little bit earlier, he was explaining evolution in terms of a random walk. I want to pick on this word 'random', which is a very slippery concept. It is negative and it is equivocal. It is negative in the sense that for a thing to be random it must be inexplicable. But the concept of explanation is itself equivocal and ambiguous; there are many different sorts of explanation, and you want to know the sort of explanation that someone has in mind before you are able to say what is random or not. Take the example of a pin-table (see diagram). There are lots of pins (B), a supply of bagatelle balls (A), and then at the bottom, a number of different slots into which they can fall (C). We start dropping bagatelle balls in. This is, in one sense, a model of a random process; but in

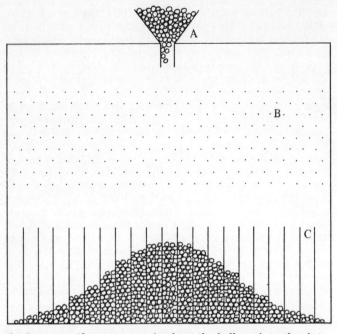

the long run, if you start seeing how the balls go into the slots, you find that the distribution is a Gaussian distribution, e to the minus x squared. Now there are two different questions we could ask about this curve. One suggestion is that it has got this shape because a small boy, like a Maxwell demon, followed each ball all the way through and put it into one of these slots. If you are asking with that sort of explanation in mind 'Is it random?', the answer is 'Yes, it is random'. But you could ask 'Is this a random distribution?' in a second sense. You are asking this with respect to the global features of the whole structure; you are asking it with various other possible explanations that you might have in mind, to see whether it is something special about this e to the minus x squared, or whether it could have been an e to the minus x, cosh x, or e to the minus x to the power of 4, or various other things. Could it have been one of those? And then the probability merchant will say: 'No, it couldn't have been one of those; it had to be e to the minus x squared'. That is, it is not random with regard to that sort of question, and that sort of explanation. And this is the point that I want to raise on the apparent randomness of the random walk which has led us here. Yes, of course it is random in one way, that is it was not carefully contrived. Later, I shall give a philosophical, even a theological, justification of why it should not have been contrived (pp. 123–30). But in another sense, it's not random; as Christopher himself said after he had introduced the Central Dogma, evolution was pretty well bound to produce something interesting. And on that point I shall stop; this, I think, is where our dispute will lie, and this is why it is going to be possible to accept the randomness of the random walk in one sense without being committed to complete randomness in another. To borrow from the title of a well-known book published in the last year, it is random in the sense that it is not one sense a *necessity* that everything that has happened had to happen just as it has; but it is not random in the sense that it is quite unintelligible why it should have happened — that it is entirely in every sense a pure matter of *chance*.

LONGUET-HIGGINS

Well, I find awfully little to disagree with in what John Lucas has said. I think that he made only one mistake, namely, when he spoke of a particular distribution of the pin-balls being random. The concept of randomness cannot describe any particular situation; it describes our expectations about the average results of infinitely many trials. In fact the concept of randomness should really be interpreted as applying to processes, not to particular detailed situations. What I was trying to say was that the processes of modification which occur in the chromosomal DNA, and which give one the variation upon which evolution can then get a hold, are random. The question 'Why did that cosmic ray arrive at that particular moment to produce that particular mutation?' seems to be a silly one, and I think that you would agree that it was. In saying this I am not denying that in surveying the primaeval soup, one might have said with reasonable confidence that once some self-replicating system had got going, something interesting was virtually bound to happen; and it's for that kind of reason that people look quite optimistically these days for life in other regions of the Universe, because they think that it's more than likely that there are certain general principles which ensure that things get more and more interesting provided you don't have absolutely impossible conditions for them to contend with. So I don't really want to disagree very much with you.

WADDINGTON

I don't want to take up most of the points that Christopher has made, because I shall be speaking later about the onto-genetic development of mind, and also about its evolution. But the question of randomness in evolution has been raised and I should like to say some things about it. The word 'randomness' has been applied to several different facets of evolution, and as John Lucas said, in each context it partly applies and partly does not apply.

For instance, it is often said that changes in the chromosomal DNA are random changes. But there is an elaborate science of mutagenesis; if you read the recent numbers of any of the main journals in genetics, they are largely devoted to

11

analysing the causes and character of mutational changes. We have in this University a very famous research group, under Dr Charlotte Auerbach, engaged precisely in this study. The changes that can happen to a piece of DNA are restricted in kind; there are only a certain number of things that can happen to it, and this is definitely a limitation on the randomness of mutational events. When people say that the changes in the DNA are random, what they usually mean is not that they are random from the point of view of biochemistry, but that they have no causal connection with the environmental factors which the adult organism is going to meet, and which will exert natural selection on it. There is a real disconnection between the changes occurring in the DNA and the tests that are going to be applied to that DNA during natural selection; to speak of 'randomness of mutation' is an old-fashioned, and by now very unfortunate, way of referring to this disconnection.

Now let's consider another context. Christopher pointed out that the evolution of higher organisms is a matter of changes occurring in populations which contain very large numbers of genes, and a great amount of variation in those genes. The populations have highly heterogeneous gene pools. When, in a situation of this kind, natural selection exerts some new pressure, so that the population has to evolve to meet a new test — perhaps a new ice-age has started or something of that kind — does the population have to wait for the occurrence of a new mutation, random in the sense we have just discussed? I think the answer is that it does have to do so *if* what it needs is something which involves only one or a few genes. If it needs to change the sequence of amino acids in some particular proteins such as haemoglobin, it has got to wait until a piece of DNA turns up that will put just the right amino acid into the right place. But if the situation requires that the animals get longer legs, or can run faster, or fly better, or fulfil some other complex requirement of that kind, there will probably be dozens of different combinations of genes that can produce an adequate result. The selection pressure is operating on some activity of the organism which involves at least several tens, and possibly several hundreds, of genes. In this situation the appearance of new DNA mutations in the gene pool may be totally unnecessary. The population can probably produce the desired result from the resources already available in its highly heterogeneous gene pool.

In an earlier lecture I used the analogy that if you need to build a very simple structure, such as a prehistoric dolmen or Stonehenge, you would have to wait till chance brought your way stones of suitable size and shape. But if you wish to erect a modern concrete building, you need not worry very much about the precise sizes and shapes of the stones in the aggregate, and the fact that these also have been shaped by chance is of very minor import. Much the same argument applies to the evolution of complex functions in higher organisms; it is true to point out that the basic elements are genes which have been produced by mutations which are random in the sense of the first paragraph, but this is trivial, and gives us almost no help in understanding the interesting questions.

Finally there is a third context in which 'random' has been used. Christopher spoke of 'a random walk' through the space of possible types of organism. From a large gene pool, in which mutation is continually occurring, a very large number of different genotypes could be constructed, and these could give rise to a great variety of organisms, larger or smaller, with longer legs or shorter legs, hairy or bald, and so on. Some of this vast array of possibilities will actually have been realised in the ancestors of any present-day species, so that one can say that the species has evolved along a path through this space of possibilities. Christopher used the phrase 'a random walk' to describe this path through the possibility space. Here I think I really disagree with him. I do not think this is a random walk at all. I think the walk is very much confined by the tests that natural selection is putting on the system. For instance, once an ancestral horse has evolved to the stage of running away instead of standing its ground and fighting off an attacking wolf, it will evolve structures and behaviours suitable for this strategy. This makes it very difficult to reverse the strategy, and start evolving instead the kind of fore-legs that would make effective weapons if it tried to stand and fight. It is much more likely that the horse will evolve further along a path consistent with its previous strategy. Its progression through the space of possible forms is therefore not at all random, but on the contrary very much conditioned by what it was doing before.

KENNY

May I join my voice to the chorus* which is suggesting that there are more senses of random than were distinguished by Christopher? I would like to take up something which he said when he was partially recanting his rejection of reductionism last year (p. 2). He said that one would not expect there to be an explanation in terms of a higher science of events at a lower level of scientific explanation, and that we would not expect for instance the transfer of energy along neural pathways to have a psychological explanation.

In one sense this seems true: one would not expect there to be general laws of a psychological type governing such events. I think that is what he meant, but surely in another sense what he said must be quite false. His own reading of his paper involved a considerable amount of activity of the kind he had in mind and the explanation of that must be psychological; I mean there could be no explanation of it which did not include the counterpart of the thoughts he had and his desire to communicate them to us.

It seems to me that the same sort of point can be made about the role which he wanted to attribute to random mutations in the history of the development of the human species. I think he was consciously putting this forward as an alternative to the idea that there was some superhuman designer who caused or supervised or designed the process which ends up with human beings like us. It seems to me that it isn't necessarily in any way an alternative. Those who believe in a divine design would think that the divine design was related to the particular mutations which are in various respects random, in the same way as the psychological factors in Christopher's lecture were related to the physiologically random events involved in his opening and shutting his mouth and moving his tongue.

LONGUET-HIGGINS

Let me explain a bit more fully what I meant when I said that one wasn't going to find an explanation for the phenomena of a lower science within a higher science. I was talking about the general rather than the contingent. I meant that it would be quite misguided to look for a psychological explanation of *how*, in general, impulses travel along nerve fibres. For a *particular* occurrence of this kind one might very well find a psychological *why* explanation — when I decide to lift this pen, the pen rises into the air, and I have explained a contingent lower-level fact in terms of a higher level science — but it would be quite another matter to explain a low-level *generalisation* in terms of a higher level science; and I think that is the way I would want to wriggle out of Kenny's rope.

On Wad's point I stand here under correction, because Wad is a very distinguished student of these matters and I am a rank amateur. But I am not sure if what Wad said actually strikes at the root of any of my main arguments. Of course, if one is going to lecture on the theory of evolution, one needs a whole course of lectures, and one would have to go into the question in what sense can a new combination of existing genes be regarded as a new biological possibility, as opposed to the appearance of a new gene produced by a mutation at a particular position in the DNA. Obviously a new combination of existing genes can provide enormous power, and it is the role of sex in biology to enable new combinations to be produced and then to be selected for. Wad also pointed out, quite correctly, that the ancestral walk which I spoke of — the random walk through the space of possible forms — is only random in a sense, and I should obviously have attempted to be more precise about the meaning of the word 'random'. But I don't think I left the matter there; I remember saying that the process was subject to a very important constraint. Perhaps the one that I mentioned wasn't strong enough, that every one of one's ancestors should have been a good biological specimen, good enough to have at least lived to the age of reproduction. But all the mutations which actually occur are probably lethal, as von Neumann established in his discussion of self-replicating machines. The mutations which we actually notice are a miserably small fraction of the total number of mutations that actually occur, and in that sense there is a very heavy selection going on right from the word go, so that these mutations are not random from that point of view. I think one of the most interesting things about the structure of science as a whole, is the role which random events seem to play in the development of processes which are very far from random. For example, if we heat one end of a copper rod and cool the other, we can be absolutely certain that heat is going to flow through the rod. The reason we can be absolutely certain is that we can treat the thing mathematically as a completely random state of affairs subject only to extremely general constraints, namely, that the centre of kinetic energy is displaced from the middle of the rod towards one end rather than the other. I don't think there is any need to get schizophrenic about the fact that the world has, from one point of view, got a lot of randomness to it, and from another point of view, seems to be extraordinarily orderly, well controlled, and intelligible. My main thesis is simply that there is all the difference in the world between saying these things after the event and saying them beforehand; between reviewing human history or human evolution afterwards to see how it went, and seeing it from the other side of time as something which inevitably had to happen.

Second Lecture. Explanations of Mind

Last night, I criticised Longuet-Higgins' Central Dogma on the grounds that the concept of *randomness* was negative and equivocal. I said that to be random was to be inexplicable, and there were many different kinds of explanations. Tonight, I want to go a bit further in distinguishing and correlating two kinds of explanations.

You will remember that Longuet-Higgins said that in his youth he had been persuaded by his elders and betters to believe that there was an important distinction between 'how' and 'why' questions, and that the scientist could only answer 'how' but not 'why'; but had since abandoned this belief, with some qualifications. Tonight, I, who am neither elder nor better than Longuet-Higgins, cannot ask him to accept my say-so; but I shall merely ask to be allowed to adduce some reasons which may persuade him that there are different types of explanations, and that these explanations need not be necessarily incompatible. Indeed, this is an obligation which I inherit from last year (*The Nature of Mind,* p. 3), when I said I was a dualist in believing that there were different types of reason, and tried to maintain, partly as a long-stop for Longuet-Higgins' own arguments against reductionism, a thesis that one sort of reason could not be reduced to another.

I now want to purge myself of a suggestion he made last year (p. 118), that if I said this I was thereby putting up 'keep off' notices round the phenomenon of mind, and saying that we just could not explain minds at all. I want to show how it is perfectly possible to have explanations of the mind, and indeed explanations of different sorts, both of the mind and of other things, without their necessarily getting in each other's way. That is my second object tonight. My third object is to legitimise Professor Waddington. That is to say, I want to show that the author of the *Strategy of the Genes* and many other works is perfectly entitled to use the personal language and the functional explanations which he naturally wants to use, and that biologists are entitled to follow their own instincts in giving the explanations of the type which seem to them to be the most rational and illuminating.

We heard yesterday that the minds which we now possess derive their origin from the practical needs of our remote ancestors. Let me distinguish two different things which our ancestors may have done and which we definitely do, which could be counted as explanations: the first paradigm of explanation − the one which we found most natural when we explain ourselves — is when we explain *why* we do whatever it is that we have done, or *why* we should do something we have in mind. These explanations of actions, given characteristically in the first or the second person, have certain important characteristics. The most important thing about them is that they should be rational and seen to be rational. One has got to be able to see why these features are relevant to this proposed course of action. The second point about them is that they are seldom conclusive. There is almost always room for further considerations. It is a matter of pro and con. We know in our own experience how often we are torn by wanting to go one way and wanting to go another way, and we have to decide between them, recognising that there are weighty considerations on either side of the case — we ponder, we deliberate; and it is because of this second point that we are able fairly easily to understand other people's minds. It is temptation rather than actual experience which widens the bounds of our understanding. I understand what it is to want to murder someone because, sometimes, although I never actually have, I have myself wanted to do away with a colleague or a pupil. I find it more difficult to understand the activities of some of the later Roman Emperors, but one can try. We draw a sharp distinction between those cases where we can enter into the mind of someone else and see why he does it, and those which seem to be entirely opaque, and we just have to take it for an established fact that this type of animal emits that sort of behaviour.

The other paradigm of explanation is where we are trying to explain *how* an event happens. For this very different ideal requirements are put forward. I don't have to see why arsenic should poison someone. If poisoning is what I want to do, all I need to know is that if I do this sort of thing, namely, give the rich uncle some arsenic, then the desired result is accomplished and my inheriting his money will ensue; that is why this paradigm has very different characteristic features. It is important that it should be universalisable in a much stronger sense than when we are just simply explaining our own actions. It is not important that the features we adduce should be either rational or seem to be rational or even apparently relevant. All that is required is that there should be some reasonably sure-fire connection between bringing about those things which are means and under my control, and the subsequent occurrence of the ends that I desire. This second paradigm I call the regularity paradigm. It was first put forward by Hume, and has become the standard explication of the concept of cause. That is, the official philosophy of science agrees entirely with what Longuet-Higgins said yesterday, and is based entirely on the regularity paradigm, although occasionally with slight qualifications.

Yesterday, Longuet-Higgins allowed that, even with a

straightforward scientific explanation, some considerations of coherence and economy were to be taken into account, as well as the uniform and reliable regularity with which the laws of nature were instantiated. Even David Hume, the Arch-Priest of the regularity paradigm, in an unguarded moment pays homage to the requirement that causes should be in some respects rationally explicable, when he allows that it is one of the criteria of there being a causal connection that the cause and effect should be contiguous. But these qualifications apart, the official philosophy of science has standardised itself on the regularity paradigm; and this has generated a world view which is characteristic, but distorted. In this world view, we see how things happen merely by virtue of knowing certain general regularities, and having some initial conditions which in conjunction with those general regularities must produce the result we are trying to explain; and so when we ask why it is that we are here and that there are minds engaging in lots more mental behaviour than was happening, say, two million years ago, the official philosophy of science says that this is a question which cannot be answered. It is just mere chance, just the result of a random walk through a space of biological possibilities. We feel somewhat affronted at this. It is a question we want to ask, indeed we may not feel only affronted but somewhat shrivelled in the face of the apparent pointlessness and uncaringness of the universe, as the official philosophy of science portrays it. For our normal assessments of what is important, it is very relevant what states of mind people have; and if when we are trying to consider our own position we are told that this is totally irrelevant, then we feel, as it were, orphaned, shrivelled and lost.

The question I want to raise is how far this view of the universe depends entirely on our adopting one paradigm. To take a traditional metaphor, people often say that our philosophy and philosophical concepts provide the spectacles with which we look upon the world. I do not say that we have, as it were, two pairs of spectacles; that was Kant's solution. If you push it through, it produces a sort of philosophical schizophrenia; you sometimes see things from the standpoint of the scientist and sometimes from that of the more engaged agent. What I want to say, instead of this, is that we have bifocal spectacles, and that normally we shift quite easily and unconsciously from one focus to another, but when we try and look at the world as a whole, we have difficulty in integrating the images which the two lenses are presenting to us.

Consider the difficulties biologists feel about functional explanations. Biologists have divided loyalties. They are scientists; they stand in awe of their physical and chemical colleagues. They are afraid that if they were to give a teleological or functional, or even a psychological, explanation of why the nerves pass through a certain pathway, Christopher

18

might be cross. But biologists are human too. They are men; and they feel the force of explanations which seem to explain why we have eyes. 'Why do animals have eyes?' 'In order to see'; one would have to be very certain of one's convictions to believe that this is not an explanation one is entitled to put forward. Biologists like to put forward functional explanations, and they do so; they think these are important, not only for the reason which came out yesterday, that sometimes a functional explanation will give us the circumstances in which some more basic regularity explanation can be put forward, but for two other reasons. First, they think of these functional explanations as being somehow schemata which enable them to gather together a whole lot of features and make them into one coherent and intelligible whole; and secondly, they use them because they have a certain homoeostatic quality which is what they discover in the biological world around them. By 'homoeostatic' I mean that if circumstances were slightly different something happens in order to remedy the situation, even in the face of adventitious adverse changes of circumstance. For example, if I get rather heated in this lecture hall, then I sweat a bit more in order to keep my blood-temperature constant. That is to say, a functional explanation doesn't lie just simply on the surface; it is also an exercise in the hypothetical subjunctive mood; it suggests, not always very explicitly, the sort of thing that might happen, were circumstances to be different. To borrow a phrase of John Austin's, it is 'constitutionally iffy'. And for these two reasons, functional explanations are going to be important in biology; to go back to the points that were being made yesterday, they set the scene in which other explanations, explanations according to the causal paradigm, can be sought for, and they make some sort of rather vague predictions at the periphery of what we know. Take the first point in the antithesis that Longuet-Higgins gave us yesterday, the 'how' and the 'why'. The 'why' questions are important because the 'why' explanations, among other things, set the questions to which some scientists will be able to give the answers. To take another case besides the maintenance of body-temperature; if I am in a bright light, then my iris expands and the pupil contracts in order that the amount of light falling on the retina shall stay roughly the same. This is a functional explanation which is illuminating, and is also going to suggest a whole line of physiological research: 'How is it that the increase in the stimulus on the retina will affect certain muscles which will cause the iris to expand?'

Functional explanations in this sense are going to gather together a whole lot of different physiological phenomena and put them into a coherent pattern. They enable us to see the point of the physiological process and also to spot topics of research the physiologist can prosecute in order to fit it all into a rational pattern of explanation. Thus far, then, the 'how' and the 'why' questions are on, as it were, two different levels, and are compatible. Where they differ is in the way they generalise. This is where the difference between the person who has in mind a regularity paradigm, whom I might call at the moment a mechanist, and a person who has a functional paradigm in mind, take different standpoints. If I look at a homoeostatic mechanism from a mechanist point of view, I shall expect to be able to explain, up to a point, how it works in purely mechanical terms, and then I shall stop. This is the way we understand a thermostat. We see how the negative feed-back mechanism works, and then we are finished. A biologist, however, is less limited in what he looks for, and often will expect a whole lot of other phenomena to come into play even outside the range of what he already knows. That is to say, when he gives a functional explanation he has at the back of his mind an expectation — which may not always be fulfilled — that even if circumstances are different in some unforeseen way, one way or another the desired goal will be achieved. He is issuing a large number of undated and partially blank cheques on future research; and he tends to express this by saying that Nature is very cunning, Nature with a capital N.

We have been considering, thus far, specific cases of the functioning of organisms, and I want now to illustrate the difference between the approaches with regard to the course of evolutionary development, in order to pick up yesterday's theme and to toss a catch to Waddington. I will take one of the examples from last year's Gifford Lectures where he described (*The Nature of Mind*, p. 128) how the horse developed the way it did, because its ancestors made a choice. They chose an environment, as much as being formed by their environment. They chose to start eating grass instead of shrubs, and to deal with predators not by turning and kicking them with their front legs like giraffes, nor by getting heavier and heavier like buffaloes, so as to be able to charge their predators out of the way, but by running; and this choice being made, a certain line of evolution is thereby going to be adopted. The horse starts by being about as big as a dog (*Eohippus*), and then grows longer and longer legs and so gradually evolves into the breed of horse we have now. Another example comes from Sir Alister Hardy's set of Gifford Lectures (*The Living Stream*, London, 1965, p. 170). Tits, as some of you are probably all too well aware, have learnt how to peck through the aluminium foil of milk bottles and help themselves to the creamy top. This has happened in our own time; it is a piece of evolution we can observe now. It is largely learnt; one tit imitates another, and Hardy argues that it gives a different complexion to our whole picture of evolution, for it is now apparent how enterprise on the part of one or two birds is copied by their friends and rivals, and produces a new lot of behaviour, and results in a new set of environmental conditions being important and a new set of evolutionary pressures operating. If you are a tit and are going to peck bottle tops, then you have got to be able to recognise aluminium foil, your beak has got to be toughened up to get through it; if, as is conceivable, the dairyman retaliates by putting on thicker and thicker foil, then the advantage will go to the tits who have stronger and stronger beaks, and perhaps soon they will be able even to get through the plastic caps that some milkmen put on milk bottles.

I am not competent to talk about the biological moral of this, but I want to draw a philosophical one. Neither Waddington nor Sir Alister Hardy are anti-Darwinians. On the contrary, they are ultra-Darwinians. What they are doing is what very often happens in the natural sciences; they are making a reverse take-over bid for the truths that Lamarck had discovered, to incorporate them into Darwinian theory. They are showing how, if we look at the course of evolution as a whole, we keep on seeing new ways which we hadn't anticipated and which don't really lie within the compass of what we should have expected, if we believed only the Central Dogma we heard yesterday. One way or another, organisms develop new ways of doing things which are going to be good for them.

To parallel this with another sort of explanation which might be given, I said 'why do we have eyes?', and answered 'to see with'. It is a problem which biologists often face. Why have very different organisms who haven't inherited their eyes from a common ancestor developed very comparable organs? The sort of answer that I like to give is to say that the earth's atmosphere is opaque to all forms of electro-magnetic radiation except for the visible spectrum and for wireless waves. Wireless waves are no good for direction-finding or for detecting things at a distance, because they are too long; they just would be confusing. Therefore, if you are going to do anything of this sort, you are going to need to have eyes which will respond to the visible spectrum. This much granted, then the rest of the story follows. We can understand the different ways insects and mammals have developed their eyes, but it follows not so much from the chances of evolutionary development but from the over-arching conditions — the way in which the pinboard (p. 10) is set up.

It is always possible to reject this sort of explanation altogether. One reason why it should not be rejected is that biologists have tried to reject it and failed. Biologists, you see, are not only human, but being human have a strong sense of sin. They feel they want to use functional explanations, therefore they feel they ought not to use them; and have often feared that if they used any such explanation they would be thought to be vitalists or accused of worshipping entelechies or something of the sort. Nevertheless, they keep on coming back again and again, offering functional explanations, with a long stream of apologetic talk pleading forgiveness for doing it, but still doing it. This seems one reason why we should accept these explanations, not only as being obviously illuminating, but as being in some sense necessary. But I can see that a person could remain blind to these blandishments. It is always possible, just possible, logically possible, to refuse to allow this type of explanation at all. It is the opposite case to that of the vitalists who refused to allow that there were any casual explanations of certain particular parts of biological behaviour, and just simply said that it was due to an entelechy. In each case, the position is tenable, logically tenable, but only just; and I should say that just as the evidence of the development of physiology provides a powerful argument against the vitalists, so the evidence of ethology provides a powerful argument against those who want to reject utterly the use of rational explanations in biology. This brings us back to where we were yesterday, and allows us to raise the question about the whole process of evolution, which seems to have its own rationale. Constantly there seem to be developing more and more complicated organisms, organisms which are in one way or another independent of their environment, organisms which are homeostatic and come to be autonomous. We want not merely to raise the question in terms of the consequences of some vast Schrödinger equation which represents the whole earth irradiated by the sun, but to ask 'why does this happen?' We are impelled to ask the question 'why?', and if we are told that this is a question which cannot be asked within the limitations of the regularity paradigm, we are entitled to conclude from this that it is showing us the limitations of the regularity paradigm rather than that the question, or any answer that we might give, is necessarily inappropriate.

This is what we ask. But in asking this question, we needn't alarm the scientists, who are concerned, quite properly, with answering 'how' questions; for, as I've tried to show, the fact that 'why' questions can be asked and may perhaps be answered, does not in the least bit show that the 'how' questions are in any way to be ruled out of court or to be regarded as redundant.

Discussion
WADDINGTON
I do not want to question the main points made by John Lucas. After all, he said he wanted to legitimise me and the kind of arguments I shall be putting forward in my lecture. At my age, being handed a certificate of virginal purity is an acceptable if unexpected experience, even if it only applies in the domain of higher philosophical thought. And, of course, I quite agree with John that it is entirely legitimate to answer

22

'why' questions about living organisms by referring to their goals or ends. This is so because the theory of natural selection allows us to see how organisms can be moulded or programmed to act in ways which would tend to bring about a certain state of affairs at a later time, that is to say to reach a certain goal. Under these circumstances the concept of a goal escapes the objections which makes the Aristotelian idea of teleology or finalism unacceptable. Many years ago I suggested that, within the realm of things subject to natural selection, the idea of a goal is only 'quasi-finalistic'; more recently it has become usual to refer to it, in a neater but perhaps less easily interpretable way, as 'teleonomic' rather than teleological.

I should like to raise the question whether we should not accept that there is a third type of causation operative in biology, which we might refer to as the historical, or, as I should prefer to call it, the epigenetic. Christopher referred to this briefly in his talk yesterday, when he mentioned the old biological slogan that ontogeny recapitulates phylogeny; translated into plain language, this means that the development of an individual repeats, in some form or other, evolutionary changes that have taken place in the ancestry of that individual. It implies that the cause for the appearance of some organ or some process during the development of the present-day individual, is to be found in the occurrence of that organ or process in its ancestors.

For the benefit of the non-biologists perhaps I should give a simple example. During the evolutionary history of the vertebrates, from the earliest primitive fish to the most recently evolved mammals such as ourselves, three types of kidney have been developed. The primitive fish operated only with the first type of kidney, and this was the organ which filtered waste materials out of their blood stream into urine, which could be excreted to the outside world. By the time evolution had proceeded far enough for the appearance of birds, this function of purifying the blood had been taken over by a second type of kidney, which is presumably more efficient than the first. We ourselves run mainly on a still later, third type of kidney. However, I want to mention only what happens in the birds.

Although by the time a young bird has hatched from the shell and starts excreting its waste products, it is operating by means of second-type kidneys, at a much earlier stage during its development the first-type kidneys put in an appearance. Why should this be so? Well, it can be shown that although these first-type kidneys never actually filter anything out of the blood stream, they do nevertheless fulfil a useful function during development. From that kidney there grows out a tube or duct, which in the fish is used to conduct the urine from the kidney to the exterior. In the embryo of the bird this duct acts as a stimulus or inducer which produces the formation of

23

the second-type kidney in its neighbourhood. Now, it is a very common feature of embryonic development that when some organ needs to be produced in a definite location, some neighbouring structure which is already in the right place is used as a stimulus to induce it. So the bird embryo probably needs something to induce its second-type kidneys to develop in the right region. But we know that quite a lot of things could do so; for instance parts of the spinal cord which are not too far away. We have to conclude that the reason why the bird embryo uses the first kidney duct to perform this function, is just because its ancestors had developed first kidneys and ducts in order to carry out the necessary process of purifying their blood.

Here we seem to have three types of causation involved. The bird embryo forms a kidney for what John calls a 'rational' reason — in order to attain the goal of keeping poisons out of its blood — which has been built into it by natural selection. Then the duct of the first kidney plays a straightforward role in 'regularity' causality, by acting as the cause for the appearance of the second-type kidneys. But the reason why it is this duct and not something else which acts as that cause, requires a different type of explanation; it does so because the bird embryo is an evolutionary modification of something in which a duct of that kind happened to be present, all ready to be pressed into service in this way.

Questions of this third kind only arise when there is some over-riding necessity which prescribes a goal to be reached, and to that extent they are perhaps nearer to John's second, rationality paradigm, than to the first, regularity, one. But I think they are different enough to require a category of their own. The sort of problem they raise reminds me of the story of the man who drove his car into Buccleuch Place, then stopped and asked a local inhabitant how to get to Tollcross. After a bit of thought about the various one-way traffic systems, and so on, in the way, the answer came, 'Well now, if I was wanting to go to Tollcross, I would'na be starting from here'.

I should now like to turn to a second comment. John has spoken of his regularity and rationality paradigms as quite distinct and separate. Are they really as distinct as all that? I tend to think that they are distinctions of emphasis rather than of kind. I believe that one cannot operate wholly within the regularity paradigm of cause and effect without having used the rationality paradigm to get into it in the first place, and vice versa.

My basic reason for thinking in this way is that we are concerned wholly with the realm of human knowledge and understanding. I do not believe that we have any access to a totally objective world which is completely independent of us. Nothing can enter into our knowledge — into the arena of our discourse — which is independent of our nature, since it is through our nature — our sense organs and general perceptive apparatus — that we become aware of it in the first place.

I know that it was convention, at least until recently, to suppose that the methods of science, following a regularity paradigm, revealed to us the objective truth; but in recent years we have seen so many radical changes of view, standing the previously accepted objective truth on its head, that I think this can scarcely be upheld any more. I suppose the classical case when science was supposed to have reached the objective truth was the discovery of Newton's Laws of Motion. You may remember the confident couplet written by Pope as an epitaph to Newton:

'Nature and nature's laws lay hid in night;
God said: "Let Newton be!" and all was light.'

But then a couple of centuries later there came the amendment by J. C. Squire:

'It did not last, the Devil, howling "Ho!
Let Einstein be!" restored the status quo.'

In order to operate within the regularity paradigm we have to use defined concepts. What I should argue is that the concepts we choose to define are profoundly influenced by the ways in which rationality principles, dependent ultimately on our telenomic goals, have led us to envisage or conceptualise the world. Something can only enter into our knowledge when we can recognise it as having a definite character. That means, in effect, perceiving that its character is near enough to some standard for it to be assimilated into that standard, and allowed to take its name. This is a process of the same kind as the assimilation of an activity under the heading of a goal.

LUCAS

Well, I very largely agree with that. I certainly want to say that there are more than just the two paradigms that I brought forward. After all, Aristotle said there were four causes (or becauses), and surely by now we can do better than that. To take another example, the whole range of mathematical explanation is different from either of the two that I have mentioned.

I'm not so happy with the kidneys being so chreodic. It seems to me that this sort of epigenetic explanation is somewhat comparable to a great deal of what we have in history. Often the skill required is not in looking for further and more profound paradigms of explanation, but rather in fitting together various bits; it is a jigsaw exercise. 'Well, there's that tube there, and what would have been here? Could there have been a membrane there? No, the spinal column would have been in a different position then.' This seems to me to be an important sort of explanation.

Explanations often satisfy more than one paradigm. Kant and Professor Hare lay great stress on universalisability as a mark of rational explanation: if I explain why I do something or why you should do something, then there is some idea that this explanation should also hold good in all other cases. To this extent, the rationality paradigm is like the regularity paradigm. Or to go the other way, I entirely agree with Wad that even if we are giving a scientific explanation, regularity by itself isn't enough; there must be some sort of rational appeal too. I agree on both these points; I only want to pick out different facets which we need to emphasise in different ways. They are potentially incompatible, and if we try and generalise too far, then we get difficulties. As long as we remain practical men, we manage to use our bifocal spectacles quite easily; but when we try and survey the universe as a whole, unless we are rather careful, we shall be caught out by these two ways of emphasising in mutually incompatible directions.

LONGUET-HIGGINS

I agree with most of Wad's comments on John's talk. John said a number of things which I certainly would agree with, but he attributed to me a number of opinions which I wouldn't want to express, in particular, the suggestion that the scientist, and especially the biologist, should reject or try to avoid using functional explanations. May I say how I see the relations between these different explanations of why we are standing here talking, using these funny things we call minds? One would want to say that our mental activity rests, among other things, upon the fact that we have brains of a particular structure. Our brains have this structure because our genes are the way they are, the message in the blueprint says 'build it that way'. But the question now arises: 'Why is the blueprint written that way?', and here we come to what Wad calls the epigenetic explanation, in terms of the way things have evolved historically. Now the question: 'Why did they evolve historically that way?', can be answered up to a point, but only up to a point; once we have decided, as it were, to try and outwit our predators and competitors, then our genetic pool will naturally drift in the direction of programs which enable us to do that better. That's what I meant yesterday by saying that all organisms tend to get better and better at what they're already good at; and so far I'm more or less in line with John Lucas.

The point at which I would possibly depart from him would be in answering the question: 'Supposing we were to go back to the year Dot, could we have foreseen any of this? Would it have been possible for anybody to say, "Aha, this must happen!"?' Here I have to declare a complete agnosticism, because there would have been nothing to help us at all in the properties of atoms and molecules as such, and as there were no organisms around at that time there would have been nothing for them to be good at or to get better at; we would be reduced to saying 'It is going to be one gigantic lottery'.

From that point onwards, of course, as the lottery turns out, goals are developed and become selected for, and the thing acquires more and more meaning and momentum as it goes on. And now perhaps the greatest component of our evolutionary momentum is our own plans, in the ordinary sense of the word 'plan'. In short, I wouldn't dream of suggesting that the humble 'how' questions are the only interesting ones; but they must be attended to if we are to understand not only the possibilities but also the limitiations of biological evolution.

KENNY

With regard to teleology, I'd like to adopt a position some-where between John's and Wad's. It seems to me that Wad doesn't allow quite enough scope to teleological explanation and John allows rather too much.

Wad says that it is all right nowadays to use teleological explanations because Darwin has shown us how to do so without lapsing into Aristotelian entelechies. I think that one has to distinguish between teleological explanations of the activities of orgahisms and teleological explanations of the existence of organisms. To take John's example of sweating, there is undoubtedly a teleological explanation of why I am sweating: it is in order to regulate the temperature of my body. Darwin comes in to point out that sweaty animals of a certain kind do better in the struggle for life than non-sweaty animals of the same kind, and that explains how there happen to be lots of sweaty such animals around and not many non-sweaty such animals. But that is not directly connected with why I am here and now sweating. That is to say, the Darwinian explanation is not in direct competition with the Aristotelian explanation. Aristotle, of course, had no explanation of the origin of species since he thought all species had been eternally there. The Aristotelian type of explanation was a teleological explanatiori of the activity of organisms, not of their origin, or the origin of the species to which they belonged.

However, I want to take up a position which is much closer to Wad's than it is to John's, because I think that when John says that there may be a teleological explanation of the whole course of evolution, he allows teleological explanation a greater scope than he has justified.

The teleological explanations which we can give are given on two different bases. Many of them, including all those which we give for the activities of non-conscious agents, are based on regularities. Therefore, the contrast John drew between regularity-type explanation and teleological explana-tion was already misleading. For instance, it is not enough that an animal or a bird should do something which is beneficial to it, for us to accept that the animal did it in order to achieve that beneficial goal. Suppose that some migrant bird, at the time of year when it is due to migrate southwards, accidentally flies through the porthole of the *Queen Elizabeth* as it is about to take off for a Caribbean cruise. This may have an effect on it which is beneficial to the organism, indeed, it may achieve a goal which we know to be one of the goals that this organism has; but not even John, I think, is going to say that the bird flew in through the porthole *for this purpose*. When we do give teleological explanations of the migrations of birds, it is because of the regularity of these migrations.

Now, of course, in teleological explanations of the actions of human beings, we don't in the same way depend on regularities. Still thinking of movements on board ship, let us change from a migrant bird to the Emperor Nero. The Emperor Nero, you may remember, brought his mother, Agrippina, on board a ship in the hope that she would drown. He was getting rather bored with Agrippina, he had had the ship cunningly constructed so that it would founder, and he not unreasonably expected that she would drown though, being a strong swimmer like Bertrand Russell in a later age, she swam to shore. Now, the reason why we know that Nero put his mother on board ship in order to drown her, is not that there is any law of nature that whenever human beings put their mothers on ships this is always what it is for. On the contrary, so far as I know, it was a unique occurrence. The reason we know that this was why Nero did it, was that he was rash enough to tell some of his friends who let it out to one of those indiscreet writers, Suetonius or Tacitus.

So this is the second way in which we discover teleological explanations: through articulate communication and language. The reason for introducing these examples is to bring out that it isn't at all clear that either of these methods of discovering teleological explanations can be applied to the course of history of the universe as a whole. We haven't any laws of nature which say that ninety-nine per cent of worlds like ours sooner or later evolve human beings, and we don't claim – or at least I don't claim, I can't speak for John – to be in verbal communication with anyone who can tell us what the plan of the universe is.

LUCAS

But I was careful not to stress teleological explanation. Functional explanations are often classified as a type of teleological explanation and teleological explanations are indeed, a type of rational explanation. Aristotle thought that they were the paradigm type but he was wrong. This, of course, doesn't dispose of Tony's point, but I suspect that when he heard me say teleological he thought that I was getting ready to rehabilitate the Argument from Design. Well, I was only getting ready. His point would be fair if I was trying to bring forward a full teleological argument; I would then have to go a great deal further, the biological argument wouldn't be enough. Kant, when he produces his critique of teleological judgement, rests it ultimately not on any crucial piece of evidence, but upon the existence of moral consciousness, on the fact that men know that they can make moral choices; and I don't suppose that I shall be able to get by with fewer premises than Kant.

I want also to turn to Christopher, and express some sorrow that in the course of the last twenty-four and a half hours, he has become more reductionist than he was yesterday. He was making the point that I can't give a detailed prediction of the epigenesis of the kidneys, or of the whole universe; this is an important point to which I shall return. But yesterday, although I allowed that we couldn't make any detailed predictions or indeed any predictions — except that evolution was pretty well bound to produce something interesting — I said that there were two different types of explanation and these answer different questions. I want to qualify this now by making two further points; that each type of explanation uses its own concepts and will phrase the question in its own terms which are not always the terms that some other type of explanation would use, and that each type of explanation is not going to be able to give us all the answers. Thus to take up the exact point that Christopher was making against me, even if I produced lots and lots of paradigms of explanations, it doesn't mean that there aren't going to be unanswerable questions. There are, and I can't give an explanation, in the terms which he poses, of why some particular sequence of events, as described by him, should happen; for instance, of why we should be speaking here. This is something which couldn't be explained in terms of any sort of regularity paradigm, nor could it have been predicted at the beginning of time. It is of crucial importance as we look at explanations to see what are the terms in which they are posed. When Christopher made his one prediction, he used a quite different set of concepts from those which you would expect a person to use if he was concerned solely with the regularity type of explanation. 'Something interesting', he said, 'would happen.' I would like to end by saying this: we can't expect there to be any sort of explanation of *all* the features which we see around us; but we can hope for some sort of *rational* explanation of why *something interesting* should be going on.

C. H. WADDINGTON

Third Lecture. The Development of Mind

I am going to talk about the development of mind, using 'development' in only one of the two senses which John and Christopher have given it, during the life-time of an individual; what is technically called the ontogeny of mind. I shall go on to the other sense of development, namely evolution over many life-times, in my next lecture, so when I use the word development this evening I shall always be meaning development during one life-time.

The development of mental activities in a baby has been studied by a very large number of people, and there have been a few studies on other animals. We have several distinguished developmental psychologists in this University. Let me assure them that I am not going to try to teach them their own business. My task is more modest. If the four of us are going to discuss the nature of mind at all thoroughly, at some point we must look at what is known about its development. My job is to draw attention to some of the salient features that are emerging from these studies, to feed them in as grist to be processed by the powerful intellectual mills of my colleagues and to be thought about by the audience. I shall only have time to give a very sketchy outline, and I shall be doing this not with inside knowledge, but as someone in a different but related field, namely the development of the material structure of the body as contrasted with its mental activities. I think there may be a number of points in which phenomena occurring in the general development of man's body may be suggestive in relation to the development of his mind. I mean not only that, if we knew a great deal more about the establishment of connections between the nerve cells in his brain, we should undoubtedly come across things closely related to the development of mental abilities; I hope to show that the character of the developmental processes occurring in many other parts of the body besides the brain may be relevant in a more general way.

One further preliminary remark; many of the most important stages in the development of mentality go on before the child is able to talk, or certainly before it is able to think critically about its own mental processes and describe them to others. Much of the study of developing mental abilities has therefore to be based on observation of a child's activities in various situations, and deductions drawn from these.

When one looks at the literature on the development of mind, one soon comes to realise that there are very few, if any, generally accepted theories about it. There are different schools of thought, often arguing quite fiercely even about relatively specialised topics, let alone about the most general characteristics of mental development. However, I think that in some cases where rival schools have been seen as mutually exclusive alternatives, they can often be regarded as opposite ends of a spectrum of continuously varying interpretations, with intermediate views which are in some sense mixtures of the two extremes. I shall consider a number of the more obtrusive pairs of alternatives, or dimensions along which there is a spectrum of opinions, which one finds in reading the recent literature.

The first is concerned with the relative importance of internal and external factors in the development of mental ability. Is the mind of a new-born baby or other animal a complete blank, which would remain blank except for the influence of inputs from the external environment? The American behaviourist B. F. Skinner comes near to the end of the spectrum of possibilities in this direction. Right at the other end would be the possibility that all mental abilities are fully described in the genes which the child has inherited, and will duly make their appearance at some time in the life history, either before or some time after birth, without requiring any assistance from the environment other than that it should permit the individual to live and grow up. Some students of animal behaviour (ethologists) hold views near this end of the spectrum, at least in relation to the behaviour of some insects and some birds. In fact, there certainly are many examples in the animal world in which some simple and not very specific factor in the external environment will act as a 'releaser' to set going a complicated performance in the responding animal. This behaviour may be shown by animals which have been reared in isolation, or in abnormal situations, in which it was impossible for their responses to have been gradually built up by earlier environmental influences. They seem to be truly innate patterns of response.

However, I do not think any students of human development would claim that there are many such patterns in man, except at the level of simple reflex actions, which we would be rather unlikely to dignify with the name of mind. Reaction to the external environment, that is to say 'learning' in the broadest sense, certainly plays an enormously greater role in human mental development than in that of the bee, ant or some of the song birds. On the other hand, most people find it very difficult to go all the way with B. F. Skinner and accept his claim that everything is learning. He has undoubtedly shown that by appropriate teaching one can train animals to do the most unlikely things; but one must ask the question, is this the process by which they acquire mastery of doing normal things in the course of their normal lives? This seems much more doubtful. You can, for instance, train a pigeon to turn round three times and then stand on tiptoe before looking for the next pellet of food in its food tray, but one cannot apparently train a normal pigeon to carry out the sequence of behaviours which constitutes the courtship ritual of the ring dove. Behaviour of that normal kind seems to depend much more strongly on internal innate factors.

The development of the human mind therefore seems likely to come somewhere in the middle ranges of the spectrum. Undoubtedly learning from the environment is important, but most people think that internal factors are also very important. What is the nature of these internal factors?

One factor which I think everyone agrees must be of some importance is maturation. The number of cells in the human brain is still increasing at birth, and goes on doing so, at an ever diminishing rate, for some time. Moreover the insulation of nerve fibres by the deposition of myelin sheaths around them is still very incomplete in the young infant, and goes on for quite a long time after birth. Clearly one could not expect a fully functioning mind when the machinery in the central nervous system is not yet fully assembled.

There is, however, still quite a lot to argue about. In particular, does the maturation of a child have to go through a definite unalterable sequence of stages? Defining a stage is not easy and is partly a matter of convenience, but people can agree on fairly broad stages, such as being able to crawl, able to walk, able to count, realising that a ball of clay remains the same size, whether it is rounded into a fistful or squeezed out into a long thin worm, and so on. The argument is about whether these stages have to be gone through in a definite order. Piaget tends to emphasise that the stages follow one another in a definite sequence. Another very eminent developmental psychologist, Jerome Bruner, argues that the sequence is very flexible. In fact he is sometimes accused of asserting that you could teach a child anything, at any age, if you set about it in the right way. That would almost certainly be going too far; I am not certain whether Bruner has ever actually committed himself to such an extreme notion, but there must be some things which cannot be done unless some preliminary has already been accomplished. You have to learn to walk before you can run, though I am well aware that this is a doctrine which seems more convincing to elderly professors than to young students. The question is to discover which are the necessary sequences, and which are flexible. Piaget emphasises the importance of a logical sequence of developments; Bruner stresses that the child may be able to acquire a number of separate abilities not in logical sequence, and only fit them together into a logically coherent whole at a later stage.

This raises the whole problem of how things which are both

32

complex and organised are brought into existence during development. This is one of the places where it may be useful to look at what we know about the development of organised complex material structures in the development of human anatomy and physiology.

The organised complex entities which appear during the development of an embryo will go on changing after their first appearance. They are not just organised structures, they are organised pathways of change, which I have called *chreods*. The point I want to make now is that, when we study the development of the material structures to the body, we find that there are two rather different and contrasting ways in which chreods may come into existence. I do not remember having seen this point made in the material I have read about the development of the human mind, although I can't claim to have done more than sample a few of the more general works by some of the best known authors, so I thought it worth explaining.

The distinction I want to make is between chreods which make their first appearance as definite coherent statements, which will then be elaborated into a whole series of details; and on the other hand situations in which a lot of isolated unrelated bits and pieces appear here and there, and are only later co-ordinated into a chreod which hangs together as a unitary organised system. The first kind, which I would call a 'generative chreod', may be compared to a piece of music which starts with a clear statement of a melodic theme which is then developed into a concerto or fugue. The second, which I should call an 'assimilative chreod', is perhaps more like what used to happen in the jam sessions of classical jazz, when half a dozen players might start doing their own thing until something gradually gelled out and the whole orchestra was working together.

In vertebrate animals the central nervous system, which gives rise to the brain, the spinal cord, and eventually all the other nerves, starts off as a clear statement, which appears on the surface of the embryo as an area with a definite shape. It appears at a definite time, and cannot be persuaded to do so much earlier or much later. It depends for its appearance on an inducing stimulus from some other part of the embryo, rather as some of the behaviour studied by ethologists depends on the 'releaser'. But again, as in that case, the shape of the initial rudiment of the central nervous system does not depend at all closely on the inducer. It is inherent in the organ itself. One can take the cells of the inducer and mix them up into a totally disordered assemblage, and, if it still acts as an inducer, what is induced does not reflect the disorder in any way but comes out as a normal well-organised system.

Just to show what I am talking about, the diagram shows the appearance of the central nervous system and brain in the embryo of an amphibian, actually a newt, over a period of about 36 hours.

I find this sort of performance, that living things seem to be able to put on, really very astonishing. There is another example, which has always got under my skin as I just don't see how it's done. In a developing bird embryo, the limbs first appear as little swellings which then grow out into either wings or legs. Right from their very first appearance they have the character, they make the statement, if you like to put it so, that they are either wing or leg; and they do this long before any of the details are decided. If a little piece of material is

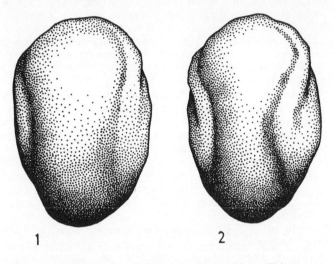

1   2

taken out of the region of the leg swelling which will later turn into the thigh, and placed in the tip of the wing swelling in a region corresponding to the toe, what does it do? It elaborates its basic theme, leg, into the detail appropriate to the region of the limb in which it finds itself, namely the toes; and when the chick grows up it has a toe growing out of the tip of its wing. It is the basic fundamental theme which is stated first, leaving the details to come along later.

But, as I said, that is not the only way in which an organised chreod can put in an appearance. You may find a number of different things appearing in different parts of the embryo, disconnected from one another at first, but later being assimilated to form a coherent system. One example would be the various organs and glands concerned with sexual activities in vertebrates. The primordial germ cells appear in one region, and they have to migrate quite a long distance through the embryo before they get into the testes or ovaries. Then the external sex organs appear somewhere else, and so do various hormone-secreting glands, such as the pituitary. It is only after they have each appeared and started doing their own particular thing, that the separate elements are woven together into a coherent system which provides the material basis for sexual activity and reproduction.

As an example, consider the little animal, *Hydra*, which is rather like a miniature sea anemone though it lives in fresh water. If you chop up a number of Hydras and mix the pieces together, you have an artificially disorganised lump, but eventually the pieces will assimilate to each other and produce one or more well organised Hydras. Another example, of an experimentally disorganised system tending to organise itself

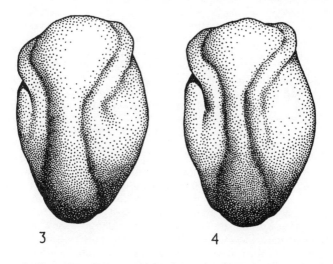

3   4

gradually into a decent chreod, can be found in the developing eye of a normal newt. An eyecup, which will become the retina, and the lens of the eye appear in the embryo. Surgically, one can substitute for the normal lens a much bigger and faster-growing one from another species of newt. One finds that the large, fast-growing lens starts growing more slowly than it should, presumably because of some effect of the eyecup; while the eyecup starts growing faster than it should, presumably because of some influence of the lens. The final result is an eye with about the normal relationship between size of eyecup and size of lens, though the whole thing is a bit bigger than it should be in this species of newt.

Between generative and assimilative chreods there is another spectrum of varying opinions about how the notions apply to the development of the human mind. At the generative end we have the well-known and important figure of Noam Chomsky; in fact I have taken the word generative, to describe one type of chreod, from his expression 'generative grammar'. He argues that the essential factor which enables the child to learn its first language at such an early age is an innate mastery of a universally applicable system of grammatical rules, which enable it to generate an infinite set of sentences, including ones which it has never heard or used before. Tony Kenny discussed his ideas in some detail in the second lecture he gave last year, and I shall not go further into them here. I will only remark that it is perhaps unfortunate, in the light of what we have been saying here, that Chomsky's ideas are discussed usually in terms of indicative or descriptive sentences. We have tended to conclude that language should be fundamentally regarded as imperative, conveying commands, programmes or instructions, rather than descriptions. Christopher has made this point, but I should like to see somebody spell out a reinterpretation of Chomsky's ideas in terms of a grammar of imperatives rather than a grammar of descriptions.

As an important figure towards, but by no means at, the other end of the spectrum one might take Jerome Bruner. One of his studies is concerned with the sucking behaviour of babies. Sucking actually involves two different kinds of action. One is an inward suck, giving rise to a negative pressure on the thing sucked; the other is an outward pushing with the lips, giving a positive pressure. When babies suck at the nipple, both these actions may produce some flow of milk with variable effectiveness. A baby normally develops a certain characteristic rhythmic alternation of positive and negative sucking. Bruner showed that if you provided it with a bottle with a teat so adjusted that it gave milk only on the negative or only on the positive sucking, then the baby very rapidly adjusted its behaviour to suit the circumstances. However, they do not remember this lesson between feeds, and at the next feed start again with their normal, presumably internally determined, rhythm. We thus have some interplay between an internal rhythm, which can be regarded as generative, and learnt behaviour, which assimilates the responses of the environment, in this case of the feeding bottle, into a modified adapted form of behaviour.

This type of adaptation is more fully expressed in another experiment carried out by Bruner, to find out how babies, one to three months old, bring together and organise into coherent behaviour two or more different types of activity. He let the babies look at a moving picture projected on to a screen, and fixed up an apparatus so that the picture only remained in focus when the baby sucked at more than a given rate. It was found that the babies soon learned to suck the picture into focus, and that they obviously like to have it in focus rather than out of focus. However, at this age they do not seem to be able to suck and look at the picture simultaneously, so they are faced with the problem how both to get the picture decently in focus, and to stop sucking long enough to concentrate on looking at it. The babies do actually succeed in assimilating these two originally diverse actions into a co-ordinated procedure which solves the problem. However, Bruner makes the important point that different babies adopt quite different strategies in achieving this. One nine-week-old baby, for instance, goes in for short bursts of sucking. Another keeps the picture in focus for much longer, but can presumably give less attention to it, while a third keeps the picture only roughly in focus by giving occasional well-spaced sucks. It is, I think, because of the variableness in the way in which these two activities are brought together into one assimilation chreod that Bruner, as we saw before, is sceptical of the idea that mental development must involve regularities of the kind suggested by Piaget's unalterable succession of stages.

In the present context Piaget stands somewhere in the middle, between those like Chomsky who emphasise the generative character of chreods, and those like Bruner who are more impressed by their possible assimilative character. Piaget admits that the formation of a new organised mental activity or behaviour does depend to some extent on interaction within whatever the environment happens to offer. That is to say it has an assimilative aspect; though he also maintains that it is much affected by internal processes of maturation in the central nervous system. He claims that when the maturation has reached the right stage the assimilation of diverse bits of experience acquired from the environment into a coherent whole proceeds very rapidly. The main point he insists on is that the character of the coherent whole formed (or schema as he calls it) is very largely dependent on its own internal logic. The external elements are fitted together in one particular way because that is the only way in which they do fit together. It is as though there is quite a short period when all the instruments in an orchestra sound a few notes independently of one another, but very quickly settle on to some melodic theme which is held together by necessary internal harmonies. Once such an organised schema has arisen it can act like a generative chreod, capable of developing into all sorts of elaborations and details within the general framework set by its basic character.

Piaget develops this idea in a direction which seems to me of fundamental importance in connection with the appearance of mental ability. He argues for a process, the title of which has been translated into English as 'Convergent Reconstructions with Overtaking'. By convergent reconstructions, he means the process by which a lot of originally divergent items of experience and behaviour become integrated into a coherent activity — the process which we have called the formation of an assimilative chreod. The point I want to draw attention to now is that referred to as 'overtaking'; the French word is *depassement*. My French is not very good, but what I think Piaget means here is something much more like 'surpassing' than 'overtaking'. He means that when a chreod is formed, it is often much more unified and coherent than is actually necessary.

This is the same sort of process as one which I have discussed in relation to evolution. It is easiest to see what is meant with a very simple example. Let us invent a model animal which is confined to two dimensions; say it floats on the surface of water, and it is assaulted from all directions by waves and wind. The best protection it has at its disposal are three stiff rods of some material, joined together to form a long structure which can lie along part of its circumference as a protection. If natural selection were to improve this protection, there would be two changes which could appear. The joints might be made stiffer and less prone to collapse; and the rods might be made longer. As soon as the sum of the lengths of the two shorter rods became more than the length of the longest rod, the structure could form a triangle. When it does, the strength of the whole structure suddenly increases many times over its former value, which depended only on the difficulty of bending the joints. The stability of the triangle may far surpass anything actually required as protection. When I discussed this notion in connection with evolution, the expression I used was that natural selection had brought into being a new *archetype*, a form which could provide the basis for a whole new range of evolutionary modifications.

As a biological example, consider the number of legs in organisms which have a hard outer skin. There are millipedes and centipedes with lots of legs, but there are not many of them, and one would not call them much of a success as groups of animals. With eight legs, there are spiders; they do a bit better, but again they are rather a minor group. When you come down to three pairs of legs, these are the insects, from many points of view by far the most successful type of animal ever produced; there are more different species of insects than of any other animal, and many of these species contain vastly more individuals than the human race. The insect form is some type of archetypal form, though nobody has any clear idea why.

Piaget argues that several human abilities depend on this type of 'hitting the jackpot', on the formation of a chroed which surpasses the needs of integration which brought it into being. In particular he uses this idea in connection with the development of logical and mathematical abilities. From a purely practical point of view, it is obviously useful to have some rough ideas about quantity, and logical argument, but man has developed mathematico-logical concepts, and rules of argument, which are incomparably more powerful than anything necessary for the day-to-day existence of a hunter or farmer, which our ancestors were.

It is not impossible, I think, that the universal generative grammar spoken of by Chomsky originated in a rather similar way; in some set of rules, which were at first favoured by natural selection because they made it possible to elaborate slightly on the degree of communication that could be carried out by ungrammatical grunts, shouts and so on, but which then turned out to be enormously more powerful than required by the immediate needs of the situation.

It seems, then, that the development of any particular mental ability involves factors of several different kinds. There are the external influences acting on a more or less recipient mentality, as emphasised by Skinner, and the internal factors, whose description is built into the genotype which the organism inherits, stressed by some ethologists; there is an internal process of maturation; and there is the final formation of an organised mental or behavioural system, a chreod, which may be attained by a generative process or by the assimilation of disorderly precursors. Probably factors of all these kinds are always involved, but with differing inportance in each particular case.

Contrast, for example, the developments of the human abilities of walking and swimming. Babies do not have to be taught much about how to walk. Well before they are strong enough to support their weight, they make movements of the limbs which foreshadow walking. When maturation, of their muscles, skeleton and nervous system, has gone far enough, they need only minimal help to start walking on their own. There is a short period of assimilating the earlier disorganised movements, then the normal walking pattern appears fully fledged, and begins to act in a generative manner; it can easily develop into running, hopping, skipping, long-jumping and so on, which use essentially the same arm-leg co-ordinations. It is difficult, but not impossible, to train people to use another system of co-ordination, such as the arm-and-leg-together style depicted in Greek vases of athletes, but on the whole the normal counter-swinging rhythm is fairly deeply embedded as an internal generative factor, and the influence of detailed inputs from the environment is small. In some other animals, such as sheep, the inbuilt factors are even more important; lambs can walk a very short time after birth.

Human swimming is in considerable contrast. We might remind ourselves that co-ordinated swimming movements are firmly incorporated into the genotypes of animals like fish or newts, which only have to mature to the appropriate stage of physical development to swim perfectly, even if they have been kept under anaesthetic from birth, and therefore have no external inputs. But it is very different with man, who normally does not start to swim until well after the period of maturation is over. He probably does not have much of an inbuilt system of appropriate co-ordinations; little more than some sort of 'dog-paddle'. The young swimmer has to do a much larger job of assimilation before he achieves anything like an effective co-ordinated activity. In this process, external factors play a much larger role in determining the final outcome than they do in walking. Some of these arise from interactions with the non-living environment; what happens when you try to breath in while your head is under water is a major influence in bringing about a suitable relation between body movements and breathing. But the major external inputs are from other people, either examples to be imitated, or teachers using speech. And a swimmer can be taught, equally easily, any number of quite different styles: breast stroke, back stroke, crawl, and so on, depending on who teaches him.

One of the most important questions about the 'higher' and more interesting human mental abilities is, how far are they like walking, based on a chreod which is rather completely specified by internal factors and susceptible to only minor modifications by environmental inputs; or how much do they resemble swimming and need to be shaped by external factors, including cultural ones? I do not think we can yet answer this at all fully about most mental abilities, such as logical or mathematical thought, anticipation of outcomes, balancing several factors against one another, and so on. But I find it easier to think that most of them are more like swimming than walking.

Even in the case which is most in the limelight nowadays, the development of speech, dare one suggest that Chomsky has allowed the undoubted importance of a generative factor to persuade him into some degree of over-emphasis? Of course his Universal Grammar only applies to the 'deep structure' of sentences, and Chomsky fully realises that from any one deep structure, many surface structures may be developed, differing not only in the actual words used but, for instance, in word order, as between English and German. But a normally sceptical reader is likely to ask whether, and how, the Universal Grammar applies to such languages as Chinese in which it looks, at first sight at least, as though the distinction between nouns and verbs is made in a different way, if it can be said to be made at all.

Then there are the claims that the grammatical structure of some languages is more appropriate to express certain kinds of philosophical outlook than others. I think it was one exponent of such views, the American anthropologist Whorf, who claimed that the only language in which it is easy to talk about the world in terms of Einsteinian space-time is one of those belonging to the Indians of the American South West, Navajo as far as I remember. Apparently they do not analyse their experiences into the separate categories of space and time in the manner we do, and their language reflects this. But if so, its grammatical structure must be considerably influenced by their cultural outlook as some very basic level, so that the generative factor cannot be all-important.

There has been at least one attempt, by a leading theoretical physicist, David Bohm, to construct a language whose grammar enshrines a Whiteheadian process of view of the nature of the world, rather than the conventional common-sense view; and the mere possibility of doing this is evidence that a deep level of a generative grammar may be modifiable by cultural factors. Chomsky's Universal Grammar may not be so independent of culture as it appears to be in most discussions of it. Possibly we need to distinguish some intermediate level, between the very deepest level which Chomsky tries to describe, and the surface. This would be concerned with the fundamental philosophical assumptions of the culture in which the language evolved. It would allow for the possibility that, in the development of language-ability, something more is involved than an innate generative universal grammar, and external factors which affect only such quite superficial matters as whether you use English, French or German words and word-orders.

Discussion
KENNY
Wad's fascinating paper was very dense, and I've only time to touch on one of the many interesting things which he had to say. Clearly much the most interesting notion which he introduced was the distinction between generative and assimilative chreods: the generative chreod in which, as in a fugue, the theme is stated first and then developed; and the assimilative chreod in which everything is much more disordered to begin with and gradually order is imposed on it. I've no doubt that this distinction is one which can be very fruitful when applied to the study of the development of mentality. However, it does seem to me that the particular application which he made of this distinction, namely, to a comparison between the approach of Chomsky and that of Bruner, was in some respects misleading.

I think that even the name is misleading if the notion of a generative chreod is meant to be generative in the same way as Chomsky's generative grammar. The notion of a generative chreod, as Wad introduced it, is a historical notion and a causal notion: in a generative chreod we have something which appears first and then directs a later stage of evolution. A generative grammar is not generative in this sense, if I've understood Chomsky rightly: the way in which the sentences of English are generated by a generative grammar is the logical sense in which all possible moves in a game of chess are generated by the rules of chess. It is a logical relation such as that between axioms and theorems, rather than a causal relation of the kind Wad illustrated for us so vividly.

I think also that Wad exaggerates the degree to which Chomsky attributes the learning of language to innate factors.

41

I'm particularly inclined to think this in view of his remarks at the end of his paper about the different cultural influences on the learning of language.

The universal grammar of Chomsky is not supposed to be, as it were, an elementary grammar of a rudimentary Esperanto which can evolve into English or French or German as the case may be. Rather, it is a set of constraints on the types of grammar which can be internalised. According to Chomsky the grammar, say of English, is a competence whose output is the performance of the English speaker. Universal grammar is, as it were, a black box, a feature of human nature which is postulated to explain how English grammar can be produced as an output from the input which is the linguistic data available to the child; what its parents and other people in the environment have to say to it. If I can use an analogy whioh for once is not drawn from computing, the universal grammar is related to particular grammars as the shape of a baking-tin is to the cake which comes out; whereas Wad thought that they were related as the ingredients, or one particularly important universal ingredient, is related to the cakes which come out.

On the contrary, it is the linguistic input to an English baby that is related to the linguistic output of an English baby as the ingredients of a gingerbread are related to a gingerbread. The universal grammar is the circular cake-tin which makes sure that the ingredients that go in come out a certain shape whether they are a gingerbread or an angel cake. For this reason, I think that Wad's fascinating distinction and the analogies which he drew from the development of the body to the development of mental activities, are misapplied if they are applied to Chomsky's theory of universal grammar.

WADDINGTON

I would like to say something about the comments Tony has made while we still have them in mind. He made two objections to my analogies, and they seem to me to be almost direct opposites of each other. At the end he talked about Chomsky's grammar as a cake-tin which simply shapes the cake that is coming out of it, but does not produce any of the cake's constituents. Now I gave, as an example of a generative chreod, a picture of the shape of the central nervous system when it first appears in the embryo. I think the analogy so far – cake-tin shaping the cake, and pear-shaped outline shaping the nervous system – is really not at all bad.

I was more impressed by Tony's earlier objection, in which he said that Chomsky's grammar is the basis for the generation of an infinite set of sentences, 'in the logical sense in which all possible moves in a game of chess are generated by the rules of chess'. Here there is certainly an important difference from what happens in development. The beginning of the central nervous system 'generates' a very complicated nervous system with an elaborate brain, all the nerves running down the spinal cord, and the peripheral nerves reaching out to the fingertips and so on; but although this is complicated it is certainly limited, and not infinite. The repertoire of a developmental chreod is always finite. If a generative grammar can generate an infinite number of sentences, then I admit the analogy breaks down at this point.

KENNY

Could I just answer quickly? I don't think universal grammars generate sentences at all. It is particular grammars which generate sentences, and universal grammar is merely a set of constraints on particular grammars. A universal grammar doesn't generate sentences any more than a cake-tin bakes cakes.

LUCAS

I would like to take up Wad's contrast between walking and swimming, and start by pouring a bit of cold water on this. About ten years ago, when my first son was born, a colleague of mine, also an eminent biologist like Wad, but committed to the doctrine of the aquatic ancestry of man, held, as one of the pieces of evidence, that babies, like Wordsworth's infants, came into the world with a capacity to swim, which they then lost − they were made to forget it by being taught to walk − and so, could he, please, borrow the baby?

These reasons apart, I'm not sure whether swimming is the best contrast to walking; but the main analogy is apt. There is one way in which we could try and push the analogy further to the direct consideration of the development of mind; and this is that in some cases the jackpot phenomenon does seem to occur. I think it is particularly so in the humanities; sometimes with poetry or with novelists or with great playwrights, there's a strong sense of 'yes, of course he's telling me what I knew before'; only, you really had to listen to someone first − Plato or Augustine, Dostoievski or Shakespeare − you didn't know it before, because nobody had actually put it in words for you. It is because the playwright is able to take you where you have already half gone, that we get the peculiar power of the literary imagination. I would like to air the possibility that this is something akin to Wad's walking. We have an ability which is already waiting to be realised; given some slight clues, you then are able to lead on much further. Not all intellectual disciplines involve this sort of ability; for example, some sorts of abstract algebras, which seem totally non-intuitive. (Not that this contrast is just that between arts and sciences. Some reasoning in the humanities is fairly opaque to the understanding and some parts of mathematics are highly intuitive.) This contrast between different sorts of intellectual abilities is like the contrast between walking and some other activities: in one case we seem to have a natural ability to develop on very slight training − we take to it very easily; in the other it comes to us only with great difficulty. I think there are

43

applications of this contrast when we consider not only our physical capacity but also our intellectual abilities; and this is one point I'd like to see developed further.

LONGUET-HIGGINS

I've just a couple of comments to make about Wad's very interesting exposition. First, I would like to put another nail in the coffin of this identification of generativity in chreods with the generativity of grammars. I don't think even generative grammars actually generate anything at all; as Tony implies, a generative grammar is more like a set of rules for playing a game: 'you can do this if you want to'. You and I generate sentences within the constraints of English grammar, and English grammar merely tells us what sorts of sentences we are allowed to generate if we want to speak a particular dialect of English. This takes me to a further point, which is that one can mislead oneself into imagining that one is explaining something by giving it a name. It's certainly very illuminating to study these epigenetic processes and to notice the distinctions between the different kinds of way in which living things organise themselves. The trouble is, one's so used to biological phenomena in ordinary life that it isn't until one sees them under rather unusual conditions like time-lapse photography that one recaptures one's sense of wonder. But I do think it's important not to imagine that one has in any sense accounted for any of these phenomena by calling them either generative or assimilative; what we are trying to do is to understand why and how these things happen.

In my first lecture I was trying to indicate the sort of statements which might commend themselves to us as explanations of living processes. If we did have a complete understanding of molecular biology we might be able to trace the unfolding of the programs which are embodied in our chromosomal DNA, and to see how those little blueprints in fact succeed in directing operations all round them and making things go this way or that, instructing this piece of tissue that it's got to get on with the job of being a wing and leaving it to the bit of tissue as to exactly how it does it. I personally think that this is by far the best type of explanation we could hope to have for the things that happen. Although, of course, we have such explanations for no more than a very, very few of the very, very simplest biological phenomena. But as we all agreed, I think, there are other ways of looking at the unfolding of the evolutionary process which brought our species into being, and these also can be illuminating because we know that there is a relation, for which we can see good reasons, between ontogeny and phylogeny; so all I would want to say here — and I'd invite Wad's comments — is that I don't think we actually explain things by naming them. I am sure that the concept of a chreod is a valuable concept, but it doesn't seem to me to constitute in any sense an explanation of the puzzles and the phenomena that we are faced with, and I only say this because I think we have an awful long way to go before we can understand the remarkable imperturbability of biological systems.

WADDINGTON

If it is true that generative grammars do not generate sentences, then don't blame me for introducing the word 'generative'. It was introduced by Chomsky, and I have only adopted it from him. And, of course, in some sense, the generative chreods I spoke about don't generate the nervous system, or anything else; they are stability constraints on the way the nervous system can generate itself, if you like. I think this discussion has got into a semantic tangle about whether 'generative' means what it says, or something else; if necessary we can return to this later.

In the few minutes left I would like to say something about the second point that Christopher made: that giving things names does not explain them. Of course not. In my next lecture I don't promise to explain the appearance of chreods, but at least I intend to talk about it. For instance, it is clear that natural selection during the processes of evolution has played a large part in building into the genotype the ability to express itself as a generative chreod, controlling the development of the nervous system, or as an assimilative one, bringing into co-ordination the pituitary, the gonads and so on. All these examples of stabilisation have been produced during evolution, and the evolutionary forces must have played a considerable part in their genesis. But one might ask the question: 'Does evolution have to do the whole job?', or are there perhaps some general rules about complex interacting systems which result in them exhibiting some inherent stability properties? I shall point out in my next lecture that there are in fact some hints in support of this second alternative. Now it is when one considers questions of this kind that I think one is going beyond the assignment of names towards the discovery of explanations. The point of giving names is to get a clearer idea of what it is you want to explain; but I entirely agree with Christopher that we should not be tempted to think that we can stop there.

Fourth Lecture. The Origin of the Soul

Towards the end of our first series of Gifford Lectures Professor Waddington complained that we had none of us given a definition of mind. To give such a definition is not in fact hard, though of course a definition by itself will not solve many problems. I want to start my contribution by offering a definition which I hope will be sufficiently broad to capture our intuitive notions of what mind should be, while sufficiently precise to provoke philosophical disagreement.

The mind is the capacity to acquire intellectual abilities. This definition, it will at once occur to you, does not get us very far unless we add to it a definition of 'intellectual'. I will do that in a moment, but first I want to draw your attention to some features of the definition as it stands. I define mind as a capacity, not as an activity. Thus it is possible to say that babies have minds even though they do not yet display intellectual activities of the appropriate kind. Secondly, mind is not only a capacity, but a capacity for capacities. Knowledge of a language such as English is itself a capacity or ability: an ability whose exercise is the speaking, understanding, reading of English. To have a mind is to have a capacity at a further remove from actualisation: to have the capacity to acquire such abilities as a knowledge of English.

By intellectual activities I mean activities involving operation with symbols. On this definition, clearly the use of language, the solution of a mathematical problem, the painting of a portrait are, as we would expect, intellectual activities. The definition leaves many things uncertain at the borderline, but this is all to the good, for so does our ordinary concept of mind that we are trying to capture in a definition. We are left in doubt by this definition whether Washoe and Sarah have minds, because we are in doubt whether they are using language, and therefore whether they have acquired the capacity for intellectual activity; and therefore whether they possess the capacity, for acquiring such abilities, which is what I am calling mind. On the other hand all the things which are well within the realm of what we regard as mental activities — like physics, philosophy, and poetry — are well within the bounds laid down by this definition. (Music is an exception to this, as to many other generalisations about the arts and the mind.) And according to this definition, I would want to say, computers very definitely do not have minds. They do, in a sense, operate with symbols, but with *our* symbols; they are not symbols for them, it is we and not computers who confer the meaning on the symbols. So I have to add something to the definition. To have a mind is to have the capacity to acquire the ability to operate with symbols in such a way that it is one's own activity that makes them symbols and confers meaning on them.

Not only does the definition have a broad borderline, it is in a way infinitely open. For many different things are called operating with symbols, and we cannot set any bounds to the possibility of the invention of new types of symbolic operation. At no time can we draw a boundary around the forms of symbolic activity current at that time and say: 'Those activities, and no others, count as intelligent activity'. Nor can we be sure that we shall ever be able to isolate from them some common element and say: 'Anything which is intelligence must have this element', unless we leave the notion of element as open as the notions of 'symbol', 'meaning' and 'activity' as I am using them.

In these lectures we have often stressed the importance of the possession of autonomous, long-term goals as a constituent of mentality. It may appear that my definition neglects this, but in fact it does not. The pursuit of self-selected goals that go beyond the immediate environment in space and time is not possible without the use of symbols for the distant, the remote, and the universal. And on the other hand, the use of symbols itself involves purposes which go beyond the temporal and spatial present. First of all, meaning something is a matter of intending, and intending involves having purposes. Secondly, to use something as a symbol and not as a tool is to use it in such a way that any effect which it may have on the environment lacks the immediacy and regularity characteristic of physical causality. So the mind, as I have defined it, is a volitional as well as a cognitive capacity: it includes the will as well as the intellect.

There are two ways in which my definition, traditional though it is, departs from familiar approaches to the definition of mentality. First, I do not take the making and using of tools as by itself an exhibition of mentality. Assisting oneself by inert instruments in the performance of an activity may or may not be a manifestation of mind; that depends, among other things, on whether the activity itself is. Thus the use of clocks to tell the time or the programming of computers to simulate natural languages is an intellectual activity; the use of a stick to shake a banana from a tree is not.

Secondly, in my definition of mind I have not said anything about consciousness. There are at least two sharply distinct things which may be meant by that slippery term. The first is the consciousness which is more or less the same thing as perception: the awareness of, and ability to respond to, changes in the environment; the senses like hearing, seeing, smelling and tasting. The second is self-consciousness: the knowledge of what one is doing and of the reason why one is doing it. In human beings self-consciousness presupposes sense-consciousness but is not identical with it. Self-consciousness presupposes something else also, I should maintain: it presupposes language; one cannot know how to talk about oneself without knowing how to talk, and one cannot think about oneself without being able to talk about oneself.

I say this last not because of any general thesis about the relationship between talking and thinking, but because of a particular reason connecting talk about oneself and thought about oneself. A dog may well think that his master is at the door: but unless a dog masters a language I cannot see how he can think that he is thinking that his master is at the door. There is nothing that the dog could do that could express the difference between the two thoughts: 'My master is at the door', and 'I think that my master is at the door'. If I am right that self-consciousness is thus intimately connected with language, then I can take account of the tradition that regards self-consciousness as closely linked with mentality without mentioning it specially in my definition. On the other hand, by distinguishing between mentality and sense-consciousness I am able to do justice both to my admiration for Descartes and my affection for my dog. I can agree with the former that animals do not have minds while according to the latter a full measure of non-mechanical consciousness.

Consciousness is sometimes identified with the acquaintance with a private world within oneself. I said several times last year that I think that this picture involves a philosophical confusion, and I won't bore you by repeating my reasons for thinking so. I will just say that the confusion seems to me to arise from people's being over-impressed with their ability to talk to themselves without making any noise, and their ability to sketch things before their mind's eye instead of on pieces of paper. I think that the acquisition of the ability to talk *about* oneself is enormously significant; the acquisition of the ability to talk *to* oneself is by comparison merely a matter of convenience. A society which differed from ours only in that everyone thought aloud all the time instead of thinking silently would be perfectly conceivable, equally intellectual, merely unbearably noisy.

My account of mind is, as I say, a very traditional one. It traces its ancestry back at least to Plato. In the tradition beginning with Plato the mind is thought of as being above all the ability to know universal ideas and eternal truths. Many philosophers have thought that Plato's belief, that beyond our symbols in mathematics and philosophy there lay such sublime entities, was due to a misunderstanding of the nature of symbolism. Perhaps so: but I hesitate to take sides definitely against Plato since the greatest philosopher of logic in modern times, Gottlob Frege, thought that Plato was basically right.

But my definition leaves the question open. If the mind is the capacity to operate with symbols, and if the way we operate symbols is correctly understood in the way that Plato and Frege understood it, then the Platonic account of the mind is to that extent correct.

In several places Plato argues, and in this he has had many followers throughout the centuries, that if the mind can know the eternal and changeless then the mind must itself be essentially eternal and changeless, and at the very least be an immortal entity that can survive the death of the body. This strand of the Platonic tradition seems to me to involve a fundamental philosophical mistake which I shall try to unravel in a moment: but one can see immediately that to get this conclusion Plato needs not only the premise that the mind is the capacity to know the eternal, but also the premise that like can only be known by like. There seems to me to be some truth in this archaic dictum if suitably reinterpreted: for example, if something is to be said to have knowledge of the multiplication table, it need not necessarily have the structure of the multiplication table (it may not even be clear what this would mean), but it must be capable of an output which is isomorphic to the multiplication table. Chomsky's argument that, since English allows unlimited embedding, no finite-state automaton could internalise English grammar would be a modern instance of the theorem that like must be known by like. But the thesis that the knower must have the same properties as the known is plausible only so far as it applies to the structural properties of the knowledge and its object; it concerns the content of knowledge and not the mode of knowing. The Platonists have never provided any good reason for thinking that there cannot be fugitive acquaintance with unchanging objects and temporary grasps of eternal truths. There seems no more reason to deny mortal knowledge of immortal verity than to deny the possibility of a picture in fireworks of the Rock of Gibraltar.

If the Platonist argument for the immortality of the mind embodies a fallacious inference, its conclusion seems to embody a conceptual confusion. If the mind is a capacity, we must ask what it is a capacity of; and the answer seems to be that it is a capacity of a certain kind of body. If this is so, then the notion of a disembodied mind is the notion of a capacity which is not anything's capacity. This seems to be as nonsensical as the notion of the Cheshire cat's smile without the Cheshire cat.

I would like to develop this by some considerations about capacities in general, in which we can hope to study the philosophical nature of the problem free of the emotional perturbances which attend discussions of survival and immortality. There are two things which a capacity must be distinguished from: its exercise, and its vehicle. Take the capacity of whisky to intoxicate. The possession of this capacity is clearly distinct from its exercise: the whisky possesses the capacity while it is standing harmlessly in the bottle, but it only begins to exercise it after being imbibed. The vehicle of this capacity to intoxicate is the alcohol that the whisky contains: it is the ingredient in virtue of which the whisky has the power to intoxicate. The vehicle of a power need not be a substantial ingredient like alcohol which can be physically separated from the possessor of the power, though it is in such cases that the distinction between a power and its vehicle is most obvious (one cannot, for example, weigh the power of whisky to intoxicate as one can weigh the alcohol it contains). Take the less exciting power which my wedding ring has of fitting on my finger. It has this power in virtue of having the size and shape it has, and size and shape are not modal, relational, potential properties in the same way as *being able to fit on my finger* is. They are not the power but the vehicle of the power. The connection between a power and its vehicle may be a necessary or a contingent one. It is a contingent matter, discovered by experiment, that alcohol is the vehicle of intoxication; but it is a conceptual truth that a round peg has the power to fit into a round hole.

Throughout the history of philosophy there has been a tendency for philosophers — especially scientifically-minded philosophers — to attempt to reduce potentialities to actualities. But there have been two different forms of reductionism, often combined and often confused, depending on whether the attempt was to reduce a power to its exercise or to its vehicle. Hume wanted to reduce powers to their exercises when he said that the distinction between a power and its exercise was wholly frivolous. Descartes wanted to reduce powers to their vehicles when he attempted to identify all the powers of bodies with their geometrical properties.

But if it is important to distinguish between powers and their exercises and vehicles it is important also not to err on the other side. A power or capacity must not be thought of as something in its own right, for instance as a shadowy actuality or insubstantial vehicle. The difference between power and its exercise or vehicle is a category difference, not a difference like that between solid and shadow. To hypostatise a power, to think of a power, say, as something which, while remaining numerically the same, might pass from one possessor to another is one way of erring on the anti-reductionist side, an error which has recently been aptly named by Michael Ayers 'transcendentalism'. In the Andersen fairy-tale the witch takes the old wife's gift of the gab and gives it to the water-butt. Less picturesque, but equally absurd, examples of transcendentalism can be found in the pages of many great philosophers.

The mind is a capacity, and the philosophical errors which

occur in dealing with capacities in general occur in a particularly vivid way with regard to the mind. Behaviourism, when it takes the extreme form of identifying mind with behaviour, is a form of exercise-reductionism: treating the complex second-order capacity, which is the mind, as if it was identical with its particular exercises in behaviour. Materialism, when it takes the extreme form of identifying mind with brain, or with the central nervous system, is a form of vehicle-reductionism: reducing my mental capacities to the structural parts and features of my body in virtue of which I possess those capacities. The Platonist belief in immortality, on the other hand, is a form of transcendentalism: for only a transcendentalist can believe that a capacity can be separated from its possessor, or pass from one possessor to another by incarnation in successive different bodies.

The title of this lecture is 'The Origin of the Soul'. Despite this I have been speaking about the mind, not about the soul; and I have been talking about its nature, without saying anything about its origin. But the points I have been making are in fact very germane to my main topic. 'Soul' is often used as equivalent to 'mind', though usually with a different emphasis to stress one or other of the mind's features; for example, to insist that it includes will as well as intellect. Sometimes it is a synonym for 'disembodied mind'; when we speak of All Souls Day it is only *disembodied* souls that are quantified. To that extent what I have been saying about minds applies without modification to souls also. In the Aristotelian tradition, however, a soul is defined as the form of the body and as the principle of life, and this definition is rather different from the one I have been using of mind.

First of all, there are many living things that do not have minds, so that if a soul is a principle of life it should belong not only to kings but also to cabbages, a conclusion which Aristotle accepted. Secondly, a principle of life doesn't seem to be the same kind of thing as a capacity for the acquisition of certain abilities. Perhaps that is only because of the difficulty of specifying what are the activities which constitute life; perhaps a principle of life is after all not much different from a capacity for metabolic change. But more seriously, if the soul is the form of the body, then it seems that a body cannot be identified as the kind of body it is unless it has the capacities in question. This seems to be perfectly correct as applied to the souls of cabbages and animals; it is a truistic consequence of being willing to accept the terminology of matter, form, and soul in their case. But in the case of human beings the soul in this sense cannot be identified with the mind: it seems that a body may be perfectly identifiable as a human body, while being unable, through some innate defect, to exhibit any intellectual activity at all.

What, since Descartes, it has been most natural to call the

mind was called by Aristotelian theologians the intellectual soul. It was this alone for which they claimed immortality (sometimes on the Platonist grounds I earlier dismissed), and this which they took to be the form of the human body. Concerning this intellectual soul they raised a number of questions which, though they strike a modern audience as bizarre and archaic, do in fact concern interesting, and live, conceptual and empirical issues. One was whether the intellectual soul was the *only* soul of the human being; another was where it came from, and whether it was created or generated. I want to discuss the second of these for a moment.

The theological debate about the origin of the individual soul began at a very early stage of ecclesiastical history. St Jerome and St Augustine held opposite views on the question. Their disagreement was dramatised by an anonymous author who uses fragments of their writing to construct a lively and ill-tempered dialogue between them, which is published with the correspondence of St Jerome in Migne's *Patrologia Latina*. St Jerome held that each individual soul was created by God; St Augustine held that each human being inherited his soul from his parents, or rather that each human soul was generated by the souls of the parents in the way that each human body was generated by the bodies of parents.

St Augustine's theory was clearly motivated by an attempt to do justice to the biblical doctrine of original sin. In one man, St Paul had said, all men sinned; and St Augustine reasonably enough thought that men could hardly inherit sinful souls from Adam unless they inherited souls from Adam; souls, like bodies, must all belong together on a family tree with its roots in the Garden of Eden. Otherwise it seems that we must think of God as somehow gratuitously stamping sinfulness on the new souls he mints afresh with each new conception of a human infant.

I don't wish to suggest that we should this evening discuss the doctrine of original sin. Some recent authors, like Iris Murdoch, have spoken sympathetically of it out of disgust with liberal optimism about human perfectibility, and some ethological writers have claimed that the human race is a uniquely murderous species by comparison with its evolutionary ancestors. But I do not think that such authors mean to suggest more than that the human nature we inherit is in some way diseased or vicious; I do not think they mean to say, as St Augustine and other theologians did, that this inheritance of a crippled nature is in some way a punishment for a historical voluntary action of a common ancestor of the race. In theological jargon, the modern versions are interested only in *peccatum originale originatum*, not in *peccatum originale originans*; in original sin but not the original original sin.

Only those who accept the doctrine of original sin in a fairly fundamentalist form have precisely the same motivation as Augustine for speaking of the inheritance of souls. The majority of scholastic theologians in fact accepted Jerome's view rather than Augustine's. Strictly speaking, those of them who were Aristotelians should have given the answer that the question about the origin of the soul was a senseless one. For the soul was the form of the body, and according to sound Aristotelian doctrine forms don't really come into existence in the way that babies do — as the Latin tag had it: *forma nec est nec fit nec generatur.*

This answer seems to me right: there can only be a question of the origin of the soul in the sense that we can seek an explanation of why human beings have the intellectual capacities they have. Later we shall have to enquire what can be said about this question with regard to the origin of the intellectual capacities of the species as a whole: at present we are interested in the ontogenetical question of the transmission of intellectual abilities from one generation to the next.

Here it seems we find in the contemporary situation something very parallel to the ancient debate between Jerome and Augustine. I am not referring to the controversy about the importance of inherited factors in the determination of individual differences in human intelligence, though this controversy has been carried on with a dogmatic fervour and a sourness of temper not too unlike that of the two Church Fathers I mentioned. I refer not to the question of whether a human being inherits a particular IQ, but how he inherits intelligence at all from the previous generation. Here, it seems to me, the disagreement which we have seen in these lectures between Christopher and John bears a more than accidental resemblance to the quarrel between St Augustine and St Jerome.

Let me explain. On the one hand it seems obvious that we inherit our soul, our mind, our capacity for the intelligent activities characteristic of human beings. Surely, it is a part of the human nature we inherit from our parents. On the other hand, John has produced a very ingenious argument, based on Gödel's incompleteness theorem, to show the human mind does not operate algorithmically. Now it seems to me that if the mind does not operate algorithmically there cannot be any algorithm for constructing it either. But according to Christopher's account last week — and indeed according to most popular presentations of the mechanisms of inheritance — human beings are constructed by algorithms. Christopher, if I remember rightly, described the chromosomal DNA as containing 'the program for constructing the human being', but if John is right there cannot be any such program. And in his dialogue 'The Seat of the Soul' Christopher suggested that the soul itself might be viewed as 'a special program in charge of all our various subroutines' (*Towards a Theoretical Biology*, 3, p. 241). So presumably Christopher, like St Augustine, takes the view that our soul is inherited from our parents. If he wanted to, he could no doubt make his system broad enough to accommodate original sin as the inheritance of a bugged program. Whereas for John, as for St Jerome, every new human being represents in some way a completely fresh start, inexplicable by the history of heredity up to that moment.

John has not, so far as I know, claimed, as St Jerome did, that each human soul was a fresh divine creation. He has offered his argument from Gödel's theorem as a proof of the freedom of the will, but not, so far as I am aware, as a proof of the existence of God. Indeed he has claimed several times that his views do not involve setting up any 'keep off' signs for science. But it seems to me that in consistency he must set up such a sign in the case of the origin of the soul. He is compelled to reject the mechanism of heredity as commonly explained. Last year he committed himself to the view, that any causal explanation must be capable of expression in an algorithm (*The Nature of Mind*, p. 74). It seems to follow that he must reject the possibility of any causal explanation of the possession of intelligence by any human being. I don't know whether this involves him in postulating a direct divine creation of the individual human soul; but if it isn't a 'keep off' sign to science, then my name is Trespassers William.

I must leave it to Waddington to say how much is metaphorical and how much literal in popular presentation of the mechanism of inheritance. I would like to learn from him whether current genetic theory does indeed imply that hereditary features are passed on by means of an algorithm. Only then will I be able to tell if the identification of Christopher with St Augustine and of John with St Jerome is as apt as it appears superficially to be.

Certainly we humans are always inclined to explain things we only imperfectly understand in terms of the most advanced technology of the age we live in. Nowhere is this more true than in the case of the mind. One of the most bizarre, as well as the most ubiquitous, misunderstandings of the nature of the mind is the picture of mind's relation to the body as that between a little man or homunculus on the one hand and a tool or instrument on the other. We smile when medieval painters represent the death of the Virgin Mary by showing a small scale-model Virgin emerging from her mouth: but basically the same idea can be found in the most unlikely places. The great Descartes was one of the first people explicitly to warn against the homunculus fallacy, but he fell into it himself when he tried to explain the relation between mind and body in terms of transactions at the pineal gland. The soul was supposed to read off images on the pineal gland, and then use the gland as a sort of tiller to steer the body on its way by means of subtle fluids called the animal spirits. The homunculus fallacy is still with us, and it is not difficult to find in the writings of distinguished contemporary

54

psychologists passages which suggest that there is a man within a man who reads the information stored in the brain and triggers off impulses which set the body in motion.

However, as time passes and technology progresses the tool or instrument which the manikin is fancied to control gets more and more sophisticated. Thus Plato thought that the soul in its relation to the body could be compared with a sailor in a boat or a charioteer holding the reins. Many centuries later Coleridge said that what poets meant by the soul was a 'being inhabiting our body and playing upon it, like a musician enclosed in an organ whose keys were placed inwards' (*Letters*, i, 278). More recently the mind has been compared to a signalman pulling the signals in his signal-box, or the telephone operator dealing with the incoming and outgoing calls in the brain. Finally, for Christopher, the boat, the chariot, the railroad and the telephone exchange give way to the electronic computer. The soul is to the body as the programmer is to the computer, so that he can describe his difficulties in handling decimal coinage as failures in programming.

I trust that Christopher means his talk of people programming their thinking as nothing more than a metaphor. As a metaphor, manikin talk may be no more than harmless necessary fancy. But it is much harder than one thinks to keep such fancies out of the constructions of one's theories, and once a manikin gets into a theory it turns into a very devil. An explanatory theory bedevilled by a homunculus is a failure as an explanatory theory: because whatever needs explaining in the behaviour of a man turns up, grinning and unexplained, in the shape of the manikin.

Discussion
LONGUET-HIGGINS
I'd like to take Tony up first of all on the things he said almost at the very end of his lecture, when he attributed to me the view that the body is to the soul as the computer is to the programmer. In 'The Seat of the Soul' I was trying to express the idea that the body is to the soul as the computer is to the programs which it is capable of implementing, and it is capable of implementing these programs by virtue of having been primed in advance with the requisite software; some very clever person actually prepares the computer so that it can implement the programs which we feed into it. Now I think it's very important in discussing the origin of the mind to make a clear distinction between the things a computer can do when it has been programmed, or a human being can do when it has developed into a mature creature, and the question of how it got that way. I think Tony has in fact rather blurred the distinction between these two issues. The question of ontogeny, as it faces the biologist in relation to the phenomenon of mind, is the question of how do the DNA blueprints

55

actually get obeyed? How did they get consulted and how did the instructions get carried out? These are very difficult and very intricate questions in developmental biology, but once the creature is mature then it seems to me we have a completely different set of questions about the relation between its mature mind and its mature body, and I would want to say that we shouldn't make the mistake of obliterating the comparison between software and hardware on the one hand, or between minds and bodies on the other, because we happen to know, as a matter of fact, that computers are put together by human beings and primed by human beings with the relevant software.

So I think I must go and ask Tony whether he wouldn't allow that there's a sense in which a successful project in artificial intelligence would, in some respects, get around the problem of the manikin or homunculus. Is it not possible in principle for a man-made system to manifest intellectual activity, or at least to manifest behaviour which we ordinarily take to be perfectly clear evidence of intellectual activity? I believe that the computer is not just another inadequate technological model of the mind, though of course it is extremely inadequate in many respects; because even if existing programs miss the mark in detail, or indeed, in important matters, I still think that we have here a very useful and important paradigm, on the basis of which we can hope to understand a little bit more about the relation between minds and bodies. So doesn't Tony allow that there is a real difference between the computer paradigm for mental activity and all previous models; and doesn't he agree with me that the origin of a system should not affect our interpretation of its behaviour when constructed? Last year indeed, he allowed that if a man from IBM arrived and said: 'I want to open you up and service you', he might be surprised, even horrified, but would not regard the incident as annihilating his own claim to consciousness or to having a mind. If, in such bizzare circumstances, Tony would resent the suggestion that he had not really been using symbols, why should he deny that a more ordinary computing system manifests mental activity when it is operating with symbols, even though we are responsible for its doing so?

LUCAS

At the moment I just want to try an *ad homunculum* argument against Tony. He defined mind as something which the body possesses; and that, of course, underlaid the main thrust of his argument to clear the swipes he was going to make against me. But later on when he was talking about his own mind, he didn't talk about a particular body possessing that mind but the little word 'I' slipped out — 'I possess', 'the mind that *I* possess' — and I think this in effect shows that his usage, as often, is better than his theory; and that we should regard minds not only as being possessed by *bodies* but as being possessed by *persons*; and that it's best to see both a body and a mind as something which characteristically is mine, yours, his, hers, theirs and so on.

I will only raise one consideration to try and persuade you that this is the right view, and the exact opposite of the mistake that Tony said was made by Platonists, transcendentalists, and for that matter by Pythagoreans, and the tellers of fairy stories, who allowed the possibility that the soul of my grandam might now inhabit a woodcock, or that the frog there might really be a beautiful prince. And it seems to me that these stories, although as far as we know untrue, and certainly very difficult to verify, are intelligible; one can conceive oneself waking up with an entirely different body; indeed, it is an important exercise in moral philosophy, particularly in the tradition of Kant and Hare, to conceive oneself not only in someone else's shoes but in somebody else's skin; and if this is intelligible, then it follows that mind shouldn't be conceived as being necessarily possessed by a body, but must be something independent of it.

WADDINGTON

I confess I feel rather out of my depth this evening. Tony defined 'mind' so that it relates only to things that use symbols, by which in practice he meant words. He defined it as essentially involving the use of language. Being a biologist who deals with the lower animals, most of my knowledge is about beings which do not have language, and therefore in these terms do not have minds. If we restrict 'mind' so that it refers only to a human characteristic, then we need some other word to fulfil a comparable function in relation to the non-language-using animals. But anyway, discussions of the meanings of words often make me feel that I am rapidly getting out of my depth, and attempts to define 'mind' and still more 'soul' soon take me into a region where I feel I can't get even one big toe down to a solid bottom. There is just one general remark I would like to make about Tony's definition. He says that mind is a capacity. Now I think that in much of modern science we have been trying to get away from speaking about capacities. Whisky has an intoxicating capacity; Benzedrin has a stimulating capacity; Barbiturates and other things have a dormitive capacity, as Molière pointed out. When I first started studying development, the orthodox answer to the question 'Why does this part of an embryo develop into the nervous system?' was that it had a neurogenic potency; and I spent quite a number of years finding facts and reasons which would persuade people that it would be more illuminating to give the answer 'Because something happens within that region to switch on the genes which synthesise the proteins characteristic of nerve cells' — or at any rate some answer with that general character. 'Capacities' always make me feel uneasy. The natural sciences do not mind talking about what Tony calls the possessor of a capacity, or its vehicles, or its exercises; it is the capacities themselves that it finds difficult to make any sense of. I should prefer to say that 'mind' is the name for a category of activities, either observed, or more usually inferred, and has a logical status very similar to that of the word 'digestion' in the phrase 'he has got a good digestion'.

Tony asked me one specific question: 'Is it legitimate to say that inheritance operates by algorithmic processes, or is that statement just a piece of fashionable jargon?' My short answer would be: 'Yes, it is legitimate, it is not just fashionable', but that needs a little expansion. Inheritance in the narrowest sense, just the passing on of hereditary factors from one individual to its offspring, is not an algorithmic, but a simple mechanical process involving such steps as the segregation of chromosomes, the union of sperm and ovum and so on. But Tony was using the word in a broader sense to refer to the inheritance of a complex characteristic of an organism from its ancestors; for instance, what is involved in the inheritance of mind? Recent biology does not say that the mind of your parent is, as it were, packed up small and stored away in some part of the egg or sperm. Nor is it good enough to say that the egg or sperm contains any thing which one might call a potency or capacity to develop into mind; what it does contain is a number of material structures (DNA genes) which specify rules for assembling amino acids into proteins. That is to say that they are programs or algorithms for carrying out some process which starts with a certain assemblage of raw materials (the amino acids) and leads to an end product (the protein) which can be defined in relatively simple terms. I think it is legitimate to say that each individual process controlled by a particular gene has an essentially algorithmic character. I do not feel quite so certain that this phrase is an adequate description of the global totality of all these part processes. In the first place all the many proteins that are produced go on interacting amongst themselves. In the second place another set of programs or algorithms put in an appearance and start operating, namely those that originate from factors in the environment. Thus the activities or characteristics of animals, or men, which might tempt us to say that they have minds are not at all straightforwardly simple readouts of DNA blueprints. They result at the end of long sequences of processes which start by operating the algorithms encoded in genes, but then go through the whole sequence of phases of interactions between these primary algorithms and between them and another set of external algorithms. I think that each individual step in this network of interactions has an algorithmic character, but I suspect that the overall result is of a kind which that phrase does not adequately suggest.

I have not yet been able to think out the implications of these ideas for John's argument derived from Gödel's theorem.

But the point I wanted to make is that in considering something like 'the inheritance of mind', I think Tony is justified in saying that we are talking about something with an algorithmic character, but I suggest we really need a theory of a different order of complexity than that with which we are used to dealing; something which is related to ordinary algorithm theory rather as population genetics is related to the genetics of individual matings.

KENNY

I'll reply to the speakers if I may in reverse order. Of course, I agree with Wad's point that it gives one no scientific information whatever to be told that opium puts people to sleep because it has a dormitive power. It seems to me, however, to be a true though uninformative statement. The reason why it's worth trying to work out the philosophical relations between capacities, vehicles and their exercises, is not because philosophy alone is ever going to solve any scientific problems about the world. The aim of philosophy is rather to help to put the questions clearly, and I think Wad is absolutely right that when one is studying capacities, what one must do is to study their exercises and their vehicles, and that all the real progress is made by correlating particular exercises with particular vehicles. But unless one distinguishes at the outset between the three, then one doesn't get clear what are the things to be correlated, and I think the history of the study of the mind confirms this. I am grateful to Wad for confirming that behind the metaphors of the blueprint and the magnetic tape, there is indeed the firm thesis that the mechanism of inheritance works algorithmically; that seems to me to present a great difficulty for John's account of the human mind, and I look forward to hearing the defence of St Jerome's direct creation theory which he promises us. But I was surprised to hear John going rather further than allegiance to St Jerome, and allying himself with the Pythagoreans. The Pythagoreans, if I remember rightly, not only thought that the soul of John's grandam might inhabit a woodcock, they thought she might inhabit a bean, and for that reason they didn't eat beans in case they might be eating their grandmother. I must say that I find the thesis that a bean might be inhabited by the soul of my grandmother not just improbable, as John says, but absolutely meaningless.

If I did say that the possessor of a mind was a body, as John says I did, then I am perfectly happy to restate all the things I've said, about the relationship between the mind and its possessor, using the word person instead of the word body, as John would prefer. For me this would not make very much difference, because I think that human persons are bodies of a certain kind. I am a body of a certain kind with certain capacities which, I argued in my paper, constitute my mind. I think that John too is a body of the same kind, and so are we all.

Finally, I should like to turn to Christopher, and I'd very much like to correct any impression I gave that there wasn't a great deal to be learnt about the nature of the mind by the analogy between minds and computers. What I was complaining about, was that there seems to be an uncertainty, if not an inconsistency, in the way in which Christopher uses the analogy. He accused me of confusing hardware with software, and I agree that it is the relationship of the programmer to the program that he often uses as an analogy, rather than the relation of the programmer to the computer itself. If it is the relation of the programmer to the program in the computer, then my complaint is that he hasn't quite decided whether the mind is the program or the programmer.

A passage that illustrates his uncertainty, his indecision, rather better than what he said last week about programming himself with regard to decimal currency, is the end of 'The Seat of the Soul' where he says: 'What kinds of thing do we really want to know about the brain? I suggest that what we would like is a detailed account, among other things, of the "software". I mean what a computer scientist would mean: the logic of the master program which sees to it that the user's program is properly translated into machine code, and implemented according to his instructions.' I don't see who the user can be here, if not the Cartesian soul. He goes on to say: 'It's quite on the cards that there is a special program in charge of all our various subroutines, which must not conflict with each other if we are to behave in an integrated way. And possibly its instructions reside in quite a small part of the brain.' The dialogue ends there, with the question: 'The seat of the soul, in fact?', but the philosophical parallel with Descartes seems at that point so close that one would not be surprised to see the seat identified as the pineal gland!

Fifth Lecture. Open Discussion

Dr Julian Davies (University of Edinburgh). *In last year's lectures, you argued that our algorithmic mechanism cannot be completely rational. Do you think (i) that human beings are not ultimately algorithmic, or (ii) that they are not completely rational? (iii) Can the human brain as a neurological mechanism fail to be algorithmic?*

LUCAS. The short answer is 'Yes, Yes, Yes'. Take the second question first — 'Do you think that human beings are not completely rational?' It is really an argument from experience. I believe there was a meeting of the Senate here yesterday, and if so the Professors here will vouch for it (if Senates in Edinburgh are anything like University meetings in Oxford), that not all human beings are completely rational. If I had to put it in the form of a formal proof, I would say that there are occasions when everybody else present disagrees with me; therefore, either I am wrong — which is absurd — or everybody else is; in which case by *modus tollendo tollens* it must be that everybody else is not completely rational; from which it follows, either way, that not all people are completely rational.

I don't think this is actually the sense in which Dr Davies was asking the question; he is putting this as an alternative to the main point of his question: 'Do you think human beings are not ultimately algorithmic?', and again the answer is 'Yes', for the reasons I gave last year; there are occasions when we can, as it were, out-do any algorithm; we can find problems which are not solvable algorithmically. Gödel's theorem shows that we could never have an algorithm which would generate just those things which are true, all of them and only those; and yet we find ourselves impelled to love the truest when we see it, and this is why it seemed to me — and I still maintain that it is characteristic of the mind — that it is in some sense autonomous and not completely algorithmic. And that raises Dr Davies' third question, which is the main point of what he wants to see discussed: 'Can the human brain as a neurological mechanism fail to be algorithmic?' Perhaps we should ask: 'How can it?'

This question becomes more of a problem in view of what we were hearing yesterday and earlier this term, when we were being invited to see the development of the human body, including the human brain, as being governed by a program; and from this it seems to Dr Davies that it must therefore be a mechanism. Now I want to concede that, subject to the qualifications which Waddington made yesterday, the many disturbances due to the environment together with the possibility of a certain amount of internal noise, the body does develop programmatically — this seems to be the best biological theory that we have — and that therefore the brain does. But to borrow a distinction which Longuet-Higgins and Kenny are very fond of using to belabour each other, the distinction between hardware and software, it seems to me that clearly all this is showing is that the *hardware*, i.e. the brain, may be of a certain definite type, but it doesn't follow that the *software* has developed as a result of a program. In some cases it may, but there are reasons why it is likely that it won't be entirely programmatic. It seems, for instance, that the software depends very largely on there being loops of neurons, and impulses going round and round and round, and a great deal of feed-back. Not nice, that is to say negative, feed-back of the sort that cyberneticians are happy to deal with, but in some cases quite possibly positive feed-back (for example, it seems a very general feature of people that one of their chief reasons for wanting to do something is they have already thought of doing it; that is, people find that the mere fact of their having decided to do something is itself a very powerful reinforcement for their persisting in doing it; in Greek Plato calls this characteristic αὐθάδεια, being pleased with one's own decisions or self-willedness, and the theologians believed it is almost a defining characteristic of human beings that they like having their own way). This is one indication of the sort of structure of the mind which makes it clear that programmatic explanations are not possible, and some very different sort of explanation is required.

At this stage, I shall pick up the point I couldn't consider yesterday which is highly relevant to the algorithms, the St Jerome versus St Augustine dispute. How far do we inherit our mental qualities through the program in the DNA from our parents?; and how far do I, like St Jerome, think that each person is completely different? As far as inheritance is concerned, of course I did not want to say, as Kenny was trying to make out, that we don't inherit mental qualities at all. It is quite clear that father and son, mother and daughter, brother and sister, very often have not only a physical similarity, but also an intellectual similarity. Sometimes you can even detect in a man's style the style of his grandfather. There is no doubt that we in that sense should be good Augustinians. But when we are talking about the mind, we are not talking only about mental qualities. The most revealing phrase is: 'to have a mind', or 'to have a mind of one's own', or 'to make up one's mind', and this is something which we don't inherit. The power of making my own decisions is not as it were, something, *just* to be programmed into me. To put it another way, if we ask: 'Do I inherit the power of making my own

62

decisions?', the answer in one sense is: 'Yes, of course, I inherit from my father and my mother the fact that I am a person, and in that sense again St Augustine is right, I have a mind of my own because I have a body of my own. Yes, this is true.' But if we ask the question: 'Do I inherit being myself?', in the sense of 'Do I inherit being me?', then the answer is 'No'. I just am me, and this is not something which I could conceivably inherit; and there I am, as Kenny wanted me to be, on the side of St Jerome.

CHAIRMAN. Dr Aaron Sloman (on leave from the University of Sussex) has some questions for Professor Longuet-Higgins, and I propose to read now the letter in which he set forth the first of these questions:

Here is my question, arising out of your apparent change of mind on the issue of reductionism. Did you mean anything more than the belief that the human mind is somehow embodied in the human brain justifies the attempt to use studies of animal behaviour to shed light on human psychology, since animal and human brain are products of the same evolutionary processes? In particular, did you also mean to imply that neurophysiological studies can be used to shed light on psychological problems?

LONGUET-HIGGINS. Well, first about my 'apparent change of mind on the issue of reductionism'. If I may say so, there is no change of mind here at all; I simply wanted to avoid being associated with the view that there's very little connection between what happens in the mind and what happens in the brain, as Dr Sloman's first question might be taken to imply. I believe that what happens in our minds is profoundly dependent on, and mediated by, what happens in our brains, and that if that were not so, there would be very little point in having brains at all. So I certainly mean – among other things – that we are well justified in attempting to use studies of animal behaviour to shed light on human psychology, and I regard the common evolutionary origin of human and animal brains as a good reason for conducting such comparative studies of animal and human behaviour.

Now for the second question: did I mean to imply also that neurophysiological studies can be used to shed light on psychological problems? My answer is, very definitely, yes. There are many psychological phenomena, especially perceptual phenomena, which can only be understood in relation to the physiological mechanisms of our senses. An example is the appearance of 'Mach bands'. In a dim light, if you look at a black and white area separated by a sharp boundary, you see in the white area, just near the edge of the black, a region which looks whiter than the rest of the white area, and this is very directly understandable in terms of the way the retina is wired up to the optic nerve. Again, colour-blindness, which is a psychological fact (I know, because I am red-green colour-blind), can be very closely correlated with the absence of certain visual pigments or their presence in only very small amounts. Or take the musical faculty of perfect pitch. This tends to go off beam when one gets into middle age, but more seriously at one end of the scale than the other; and this has a perfectly simple explanation in terms of the hardening of the cochlea. Cells which previously responded to a particular frequency now respond to a different frequency and so middle C sounds lower than it used to. So in certain matters at least, physiological considerations can shed a great deal of light on psychological facts. But, of course, the neurophysiologist cannot hope to understand how the brain works until he has equipped himself with a non-physiological account of the tasks which the brain and its various organs are able to perform. Only then can he form mature hypotheses as to how these tasks are carried out by the available hardware.

Aaron Sloman in his letter goes on to say something rather harsh about neurophysiology — he says that until much more elaborate theories about the nature of psychological processes are available, neurophysiological research is bound to be irrelevant to psychology. Probing a computer with screwdrivers would be no way to understand how a computer program managed to talk or understand English, if that is what it did. We certainly aren't going to discover how programs work by poking around with electrodes, or by detaching large chunks of computer. At the end of his letter, he says: 'I'm so far very disappointed that none of the other speakers has shown any interest in the implications of the thesis that the mind consists essentially of programs. Perhaps they don't understand it. What have biologists to say about the evolution of programs? Do they understand enough about programming and computation to be able to think about the problems?' Perhaps Wad had better field this one!

WADDINGTON. Of course I am not an expert on computers, and I am not sure that I know much about programs now that they have been turned into a professional expertise; but I have been discussing biology in terms of concepts which I think it is fair to call programs since long before computers emerged from the backrooms of laboratories. As a developmental biologist, one is concerned with controlled sequences of processes, and such questions as whether one step is controlled by the immediately preceding steps or by something else, and whether there are sub-routines which can be switched on in such a way that you can cause a cell to develop into some adult type which it would not have done if left alone. All these questions could very well be phrased in the language of programming, and I would be quite willing to use the programming language as soon as I become convinced that this adds anything to what biologists have been saying already. But I think that having new ideas, such as that of a chreod, which we could call a self-correcting set of programs, is more important than the actual language one uses to express them.

As a biologist, one is interested mainly to discover the actual programs carried out by, say, a newt's egg. One wants to find out about them both in terms of the hardware, that is what materials are being used, and also in terms of the software, by which I mean such questions as whether the subroutines can be brought in in random order, or whether they must come in a definite sequence, or whether you can call them up at particular stages of development. In my experience, the general theory of computation hasn't added much to our understanding of the things a biologist deals with, even if you regard those things as operating like computers. However, I may not know enough about programming and computation, and perhaps it really does have something to tell us; but if so I wish somebody would tell us what it is.

CHAIRMAN. Now, Dr Sloman has what is not so much a question as a critical comment, again to Professor Longuet-Higgins:

Christopher Longuet-Higgins seems to want to compare the mind with an interpreter or compiler. This seems to be an extremely restrictive analogy; the mind must be a massive multi-purpose program including problem solving, subprograms and large stores of modifiable data, some procedural.

LONGUET-HIGGINS. Well, perhaps I should just explain what an interpreter or a compiler is. An interpreter is a system which will accept symbolic instructions in programmatic form, and actually get them carried out. When you use a computing system, in one sense the whole thing is an interpreter, but it is important that the interpreter should be able to understand the language in which you address it. For this purpose a particular clever bit of software called a compiler is specially written, and this enables the computing system to make sense, in terms of its own machine code, of the instructions as they arrive in the programming language. A compiler essentially converts what you type in the high-level programming language, such as Fortran or POP–2, into machine instructions which are then implemented. I was certainly not identifying the mind with a compiler in that sense; as I said last year on p. 25 of *The Nature of Mind*: 'I want to suggest that the problem of describing the mind becomes very much clearer if we recognise that in speaking of "the mind" we are not speaking of a static or passive entity but of an enormously complex pattern of processes'.

I'm not singling out any one of those processes and saying: 'That's the mind', though some of the processes in the complex system of programs which we call the mind will be more crucial than others; in other words, there are certain programs which you can do without and still be yourself. If I were to go blind, then the programs which mediate my vision would be out of action; but the most important ones, which I could not do without, are those which really make the essential me. So I would wish to include, in any account of my mind, all the mental processes of which I am capable, and in which I engage, such as remembering, imagining, reasoning, using language, playing chess. So I would agree with Aaron Sloman that one mustn't be too restricted in one's use of the term 'mind', and if one accepts the idea that the human mind is essentially all the programs which a human brain can carry out, then we have the right to say that the brain is by far the most impressive computing system that has ever existed, or will — probably — be invented.

CHAIRMAN. Now we have another question from Dr Sloman, this time for Dr Kenny:

In your last lecture, did you not fail to see that a program is itself a new kind of manikin? In this sense: unlike previous instruments, that is, the product of previous technologies, a program can make, modify and control other programs including itself. For example, a program can call itself recursively, modifying itself with each call. It can also construct and use symbols for its own current purposes which the original programmer knows nothing about. That is, they are not the programmer's symbols.

KENNY. Here I must begin by pleading guilty to having misunderstood Longuet-Higgins. I argued yesterday that he was undecided whether to compare the mind to a program or to a programmer. I argued this on the grounds that while in his own experimental work it is the program which corresponds to the mind of the language-speaker, in his paper, 'The Seat of the Soul', and in his lecture last week, he had compared the mind to the *user* of a computer, to a programmer. He has now convinced me that I quite misunderstood the passage in 'The Seat of the Soul' which I quoted in support of the accusation that he was smuggling in a homunculus or manikin. He said there that one of the things that we would want to know about the brain, was the master program which sees to it that the user's program is properly translated into machine code and implemented, and then he went on to talk of the soul as a controlling program. Now, I took it that this definition, though of course explicitly it equates the soul with the program, had by speaking of a *user* smuggled in an illegitimate reference to what could only be a homunculus or Cartesian spirit. But I now see that the role which in the analogy is played by the user is supposed to be played in the analogate not by the soul but by such things as eyes, and the ears. Consequently, this passage doesn't involve a homunculus and I apologise for saying that it did.

Likewise, when Christopher spoke of himself as reprogramming himself to cope with decimal currency in the lecture last week, this wasn't meant as a transaction between a homunculus and a program but rather a transaction between the master program with which Christopher identifies himself and the subordinate program which contained instructions for coping with currency.

Christopher indeed claimed that the great merit of his approach is that for the first time it enables us definitely to kick the homunculus out of doors when we are trying to explain the mind. I'm not yet convinced that it does so, simply because it hasn't yet got far enough. The type of work which has actually been done in artificial intelligence hasn't yet dealt with the areas in which the homunculus is traditionally invoked, for instance, the taking of spontaneous decisions and the conversational use of language. Christopher's program can answer in English questions put to it in an appropriate form; it can give yes-no answers, but it doesn't go in for the stimulus-free use of language which Chomsky has so much emphasised as the characteristic of human beings.

Now, I think that this recantation can serve as an answer to the first part of Aaron's question, but with regard to the second part about symbolism, I'm quite unrepentant. The operations of computers upon the information and instructions which they contain are not, I would maintain, symbolic operations, even though the input to and output from a computer may be symbolic in the fullest sense; as it is, for instance, when one talks to Christopher's program in English in playing this game 'waiting for Cuthbert' which he described to us so vividly last year. When one does use English sentences on the teletype in communicating with Christopher's program, then of course, one is using symbols which have meaning and symbolic meaning, but it is a meaning which we have given them; and moreover it is the meaning which we have given them in our transactions with each other and not in our transactions with the computer. But I suspect that here there lurks a deep philosophical disagreement between Dr Sloman and myself, not on the nature of computers but on the nature of symbolism.

Mr Cavanagh (University of Edinburgh). *It is perhaps the case that among the constraints on the form of natural language are the following: (i) A linearity imposed by the physiological characteristics of the acoustic medium. (ii) The existence in the world of the physical dichotomies, motion vs state, and cause vs effect. (iii) The essentially 'binaristic' nature of much human thinking. Is this a reasonable assumption? If so, could Dr Kenny suggest what additional constraints might be imposed on the form of the grammars of natural languages by a universal grammar such as that posited by Chomsky?*

KENNY. I'm in the fortunate position of being able to give an example of the type of constraints imposed by universal grammar. Last night I was privileged to attend a brilliant presentation by Dr Edward Keenan of the work on noun-phrase accessibility and universal grammar which he has been carrying out with his associates in Cambridge. Let me explain. Keenan's work concerns the formation of relative clauses. In English we can relativise noun phrases standing in very different positions in a sentence; for instance, we can relativise things standing in subject place, as in 'the woman who is married to John'; or in direct object place, as in 'the woman that John married'; or in indirect object place, 'the woman to whom John gave a ring'; after a preposition, as in 'the woman that John is sitting next to'; after a possessive as in 'the woman whose sister John married'; or, finally, after a comparative particle, as in 'the woman that John is taller than'. There are others too.

Now, Keenan was explaining that not all languages allow relativisation in all these places. Some Malay languages, for instance, allow relativisation only in the first case and not in the others. Keenan's work suggests that these possibilities can be arranged in a hierarchical order, in fact, in the order of the previous paragraph. It suggests that we have a hierarchy thus:

$$\text{Subj} \geqslant \text{DO} \geqslant \text{IO} \geqslant \text{O–Prep.} \geqslant \text{Poss NP} \geqslant \text{O–Comp}$$

Any language which admits relativisation at any point on this scale, also admits it at any point further back towards the left, but not conversely; so that you can group languages according to where they drop out.

The hypothesis of such a hierarchy has been confirmed by the study of forty or so languages. Now it seems to me that this hypothesis provides a perfect example of the type of thing that a universal grammar is supposed to contain. The hierarchical principle will be in a principle of universal grammar, something obviously highly abstract.

Now I don't think that a constraint such as this hierarchical principle could be adequately explained by the constraints which the questioner listed. Consider the contrast between universal grammar and logic. The relevant part of logic would be the first order predicate calculus with identity. If we all spoke, instead of English, a version of first order predicate calculus which included English nouns and predicates then we would form all these relativisations by using something called the iota operator which means roughly 'the x such that'; you have 'the x such that x is a woman and x is married to John', 'the x such that x is a woman and John married x', 'the x such that x is a woman and John gave a ring to x', 'the x such that x is a woman and John is sitting next to x', 'the x such that x is a woman and John married the sister of x', and so on.

Now, there are absolutely no constraints on relativisation in the first order predicate calculus. By this I mean that wherever a noun phrase of the appropriate kind can occur at all in a sentence, in any of the positions where these xs are, then it could also be relativised. There is no hierarchical principle either; there is nothing more or less natural about relativising in the most complicated sentences than in the simple ones with which we began. But natural languages, even ones which are as generous about relativisation as English is, don't allow things which are perfectly all right in logic. For instance, in logic there is nothing wrong with this: 'the x such that x is a woman and grass is green and John married x', but even English is pretty unhappy about 'the woman that grass is green and John married'. Now, when we use logic we are of course subject to all the constraints imposed upon us by the linearity of the acoustic medium, by the contrast between concepts of motion and state, and so on, that were mentioned by the questioner. We are influenced by all of these, so we need some further explanation why there are extra constraints that operate in natural languages which don't operate in logic, and that is what Chomsky's postulation of innate universal grammar is meant to supply.

Mr Alex Solomons (University of Edinburgh). *The human brain seems to have an unlimited ability to generalise and extend concepts. During this process, the mind is thinking in terms of the then highest level of organisation, that is, the highest system appropriate to some concept. It is often tried to place the concept in a higher and as yet unknown system. The fruit of this activity is the discovery of the higher, more generalised and usually more meaningful concept. This creative act may involve the thinker in momentarily relegating the lower levels of organisation to the semi-conscious but they are still accessible to him or her. My question is this: does the form of Chomsky's transformational grammar give any insight into this process of generalisation? If the answer to this last question is yes, I would like to extend the query, to generalise my question. It would seem to me that music and the visual arts also have their universal grammars. Is one not conscious of the generalisation and the extension of concepts in the fine arts? This being so, what is the possibility of a super universal grammar covering these universal grammars, or is Chomsky's transformational grammar adequate when applied to the realm of the fine arts?*

KENNY. The question is, does Chomsky's transformational grammar throw light on the process of generalisation? Well, I think the answer is that it attempts to. Generalisation in the sense intended by the questioner, is a semantic matter. Grammar, as defined by Chomsky, includes a semantic component as well as a syntactic one, and a number of Chomsky's followers have produced theories about the semantic component of grammar. My impression is that these are not generally held to have been as successful as the syntactic studies of the transformational grammarians.

Again, in answer to the second question, there have been a number of attempts to apply techniques analogous to those of transformational generative grammars to such things as the rules governing musical composition in different musical traditions. I think that attempts have even been made to link this work with work on other species, to link it say with the work of those like Thorpe, who made studies to attempt to separate an innate and an acquired component in the song of song birds like the chaffinch. I'm not familiar with the studies nor competent in the relevant fields, so I cannot say how successful the work has been. Perhaps in a moment Waddington might say something about the latter studies. I should insist that I'm not a linguist and that when I took Wad to task the other night about Chomsky, it wasn't because I thought I was entitled to an opinion about the truth of the theory of universal grammar, but merely because I thought Wad had in some respects misrepresented the structure of the theory.

WADDINGTON. I don't think that I have very much to say on the subject of bird-song, but the situation seems to be this. Each species of bird has a more or less well-defined type of song. In many species this is completely fixed in its heredity, and you cannot persuade the birds to sing anything else. In quite a number of other species the song is more flexible; by bringing the bird up in suitable ways you can persuade it either not to sing at all, or in some cases to modify its song to some extent, or even to sing in a way characteristic of other species. I do not quite see how this links up with anything particularly universal. Possibly you might say that each species has its own 'universal' grammar. But most bird-songs are fairly simple; in many species it is just a repetition of a single phrase, and I do not see much point in discovering a universal grammar within a single phrase. Of course there are some complicated bird-songs, such as that of the nightingale which usually does not repeat itself. It would conceivably be interesting to see if you could make any sense of the idea of a universal grammar for the nightingale.

CHAIRMAN. Dr John Beloff (University of Edinburgh) has sent us a question which he addressed to Professor Longuet-Higgins, who has taken the liberty of re-routing it to Professor Waddington. I am presenting it here in somewhat shortened form:

In presenting your evolutionary approach to Mind you warned us against adopting a too intellectualist theory of Mind and asked us to remember that behind every thinker of today lies the ancestral hunter and man of action. While I would go along with that as a broad generalisation there are, I would say, some very striking facts that do not easily fit this evolu-

tionary interpretation. Consider, for example, such human activities as music, mathematics and chess. A high level of aptitude for any of these fields seems to be both fairly specific and, to a large extent, hereditary, as evidenced by the fact that it reveals itself at such an early age. Now, the question I would like to put to you is: what conceivable advantage would the possession of such abilities have conferred on individuals at the time when our species was being established?

One other point. I would like to draw your attention to a passage in Koestler's Ghost in the Machine *(pp. 272-3). There he cites several authorities, including Le Gros Clark, for the view that the enlargement of the hominid brain from the mid-Pleistocene era onwards proceeded at a pace 'exceeding the rate of evolutionary change which has so far been recorded for any anatomical character in lower animals' and that it represents a case of 'explosive evolution'. What I should like to ask is: (1) can this development be accounted for in terms of what was strictly required for survival under the circumstances of that period? and (2) does it make sense to suggest that evolution can give rise to pathological developments? Koestler mentions the idea that 'the human cortex is a sort of tumorous outgrowth that has got so big that its functions are out of control'. Do you think there is anything in this pessimistic idea?*

WADDINGTON. Dr Beloff's letter contains quite a number of questions and I shall answer it under six headings. Firstly, there is the point about 'ancestral hunter and man of action'. I would just like to remind you that our ancestry includes mothers as well as fathers. Our inheritance includes contributions from food gatherers, cooks and housekeepers as much as from the chaps who were good with the bow and arrows.

Next is the point about specialised activities, such as music, mathematics and chess. I think there is little evidence that most of these abilities are hereditary, except perhaps in the case of music. It seems to me almost certain that most of these highly specific activities depend on fortunate and relatively rare combinations of genes. They are underlain by complex genotypes, rather than the presence of single hereditary factors. I suspect they must involve something like the phenomenon which I referred to earlier as 'hitting the jack-pot', achieving a sudden large leap in efficiency just when all members of some complex set-up fit perfectly into place. In this situation, evolutionary processes will have been concerned mainly with the natural selective value of the individual components of the complex system, rather than with the complex as a whole, which would only come together in the right configuration very occasionally.

Then there is the question, raised by Koestler and Le Gros Clark, that the enlargement of the hominid brain is a case of explosive evolution. I should point out that most major transitions in evolution have gone extremely fast so that they have left very few intermediate stages behind. This was the case, for instance, in the evolution of reptiles into mammals, or into birds. I am not convinced that the evolution of the hominid brain was so much faster than these other examples.

Then Beloff asks can 'this development be accounted for in terms of what was strictly required for survival?'. In connection with this I think I should again refer to 'the jackpot effect'. There is really nothing surprising if evolution sometimes produces 'more' than is strictly 'necessary'.

Then there is the question whether evolution can give rise to pathological development? The answer is that it certainly can produce states which will appear pathological if, for instance, the environmental situation changes so that the criteria of natural selection are altered. It is quite possible for natural selection to favour a developmental process which produces a condition favourable to the animal during the period before reproduction takes place, but which goes on developing into a possibly unfavourable condition in the later stages of life, with which, of course, natural selection is not much concerned. An example of this may have been the enormous horns which were grown by the Irish Elk, which must have been a great handicap in the later stages of life, and may have led to its eventual extinction when the environment became generally less favourable.

Finally, there is Koestler's idea that the human cortex is a sort of tumorous outgrowth which has got out of control. I don't much like the idea myself. Developmentally the human cortex is a perfectly well behaved organ, with a definite size and morphology, not at all comparable to a continuously enlarging disorderly cancer. I think that the real point that Koestler is making concerns the destructive tendencies in human behaviour, which seem to go along with increase in foresight and intellectual ability, and to have evolved along with the enlargement of the cortex. Personally, I do not look to anatomy to give an explanation of this. I have advanced another argument on this subject, which I have discussed in some detail in a book called *The Ethical Animal*. I have not time to expand it fully here, but the gist of the idea is that human mental abilities depend primarily on the fact that man is a language-using animal. He has only become so because he has found some method to convert neutral stimuli (e.g. noises) into symbols with meaning (e.g. words). In practice he does so by associating noises with the control of behaviour by external authorities (e.g. parents), in the experience of the very young infant; but during this process, by which the baby is initiated into the world of culturally transmitted information, a 'hitting the jackpot' phenomenon unfortunately tends to occur, by which the internalised authority necessary to convert noises into words becomes hypertrophied into what Freud referred to as a 'super-ego', and it is this which is responsible for the destructive tendencies which Koestler attributes to an overgrowth of the cortex.

CHAIRMAN. Perhaps there is just time for Professor Waddington to make one or two remarks about a final question which comes from Dr Walter P. Kennedy (University of Edinburgh):

May I raise one point that has not been mentioned so far, though of course it may come up in one of the later lectures. This is Stephen Black's view of the development of mind in his book Mind and Body, *1969. Two brief quotations must suffice to provide a basis: 'I therefore put forward the definition "by Aristotle out of Schroedinger": that life is a quality of matter which arises from the informational content inherent in the improbability of form' (p. 46). 'I therefore put forward the second hypothesis: that mind is the informational system derived from the improbability of form inherent in the material substances of living things' (p. 56).*

WADDINGTON. I think Stephen Black is right in arguing that both life and mind are more closely connected with what he refers to as 'information traffic', than with the material interactions studied by chemistry and physics; though I should prefer to use the phrase 'instructional traffic'. We could, for instance, have systems worthy of being called alive which were built of quite different chemical materials to those used by living organisms on this earth. However, I think that when Black goes on to the second point quoted by Dr Kennedy, and identifies mind with any 'informational system derived from the improbability of form', he is really enlarging the scope of the word too much. I think that if the word 'mind' is to remain useful at all it has to be restricted to something to do with the central nervous system, and of course, there is plenty of improbability of form inherent in many other material things than that. I hope I shall have time to say something further about these points in my next lecture, on 'The Evolution of Mind'.

Sixth Lecture. The Evolution of Mind

A lecturer who takes as his subject the evolution of mind is confronted by two main challenges. He must consider how anything worthy of being called mind could have originated within a world which, we have every reason to believe, originally consisted only of non-living matter obeying strictly the laws of chemistry and physics; and he must face the problem of how mind-like activities, after they had made an initial appearance on the scene, have evolved to their present state. I will argue that two general principles can explain, or can at least throw considerable light on, both how mind originated and how it evolved. The first of these principles is the Darwinian theory of evolution by natural selection, but we shall need to understand that theory in the form in which it has been formulated in the last few years, rather than in the way which has been considered orthodox over the last three or four decades. The second depends on one of the more recent insights into the modes of interaction between things, which is often referred to by the generic title 'information', though as we shall see this is not a very happy term.

I shall not at this stage offer any definition of 'mind', taking it that we all know, roughly speaking, what we are talking about. Definitions are, in my opinion, not things to start with but things to approach. Possibly by the end of this lecture we may find ourselves in a more favourable position to offer a definition with some substance and value.

Tracing the evolution of life backward in time, by following the fossil record further and further into the past, we find the variety of living things gradually becoming reduced, the more complicated ones dropping out earlier, so that eventually we find evidence only of quite simple creatures; and at some time in the pre-Cambrian even the evidence of bacteria and algae disappears. At earlier times than that the stage on which we now operate seems to have been lifeless, composed only of chemical atoms and molecules, probably in very different concentrations from those we now know. The atmosphere, for instance, contained very little oxygen and a good deal of ammonia, and the sea was much less salt. In what terms is it reasonable to try to give an account of how living things, particularly things exhibiting minds, could have made their appearance?

There is one school of scientific thought which claims that not only do we have to start historically from an age in which nothing was happening except the chemical and physical interactions of atoms and relatively small molecules, but that also we have to accept the laws of physics and chemistry as the sole foundation for any intellectual scheme about the formation of more complicated things. This is the view nowadays usually referred to as reductionism. It argues that nothing can be considered explained until it has been reduced to physics and chemistry. Earlier in this century, this view was more usually referred to as mechanism. At that time there were still people who repudiated it in extremely drastic ways, arguing that living things involved not only physical and chemical entities, but also some sort of 'life force' or 'vital principle'. These 'vitalists', however, did not succeed in making it at all clear just what sort of thing they meant by a life force, and in general they had little influence on the development of biology. On the other hand, from about the thirties onwards, there was another group of influential theoretical biologists who maintained that there is indeed something more to living things than mere physics and chemistry. What they suggested adding to these was not any vitalistic life force, but rather various types of complex inter-relationships between the simple physico-chemical entities; and because of these interactions the complexes exhibit properties and behaviours which cannot be shown by the elements when they were not so related. The separate parts of a car can, when assembled into the right relationships, exhibit the new property of locomotion, including such (at first sight, improper) movements as moving up a hill against gravity. It was suggested that, in a comparable way, the properties we see in living things depend on their physico-chemical parts being assembled in the correct relationships. These 'organicists' therefore argued that life involves physics and chemistry 'plus organising relations' (Needham, Woodger), or 'plus systems properties' (Bertallanfy).

This is quite a useful formulation. There is no doubt that to explain the properties of living things we have to add something to the bare notions which suffice to account for the chemical and physical behaviour of atoms. However, I have always felt that the method of approach of both the reductionists and of the organicist anti-reductionists is fundamentally upside down. Our understanding of the world does not start from a firm knowledge of physico-chemical entities, to which we may or may not have to add something in addition. Following A. N. Whitehead, I should argue that any knowledge we may succeed in acquiring about the world starts from something totally different; namely, from occasions of experience. In an 'occasion of experience' there is involved both a knower and a known, a subject and an object, and the content of an occasion of experience is undoubtedly influenced by both of these. The knower has a certain apparatus for perception, and also certain pre-existing interests, some of which are innate and expressed even in the very earliest stages of life: he has what Popper has called a 'prior knowledge'. Meanwhile the known has, of course, its specific characteristics, which colour the content of the occasion of experience in so far as the knower has the perceptive equipment or the interest to discover them. According to this view the physico-chemical concepts, such as mass, force, energy, atom, etc., are not the items with which knowledge starts. Instead they are notions which we have found it useful to derive from certain occasions of experience. The basic physico-chemical notions are in the first place derived from experiences which involve rather simple contents, such as balls rolling down inclined plains, pendulums, and the combination of hydrogen and oxygen to make water, or of sodium and chlorine to make salt, and such like.

According to this view the fundamental basis of scientific knowledge is not in any set of scientific entities, such as atoms, whether Daltonian, Rutherfordian or quantum mechanical. It is in experiments, which are occasions of experience which have been organised and experienced in a systematic way. The fundamental problem is not to build up a complex phenomenon, such as mind, from a previously given set of simpler entities, such as atoms; on the contrary, it is to search downwards from the complex phenomena to simpler entities adequate to explain them. It is, therefore, not quite correct to say that we have to add 'organising relations' to the physico-chemical entities in order to explain life. It would be better to say that in explaining simple chemical behaviour, we can leave out certain properties of the elements which only become important in large and perhaps specially ordered arrangements. For instance, the classical laws of organic chemistry, and the classical descriptions of chemical elements and small molecules, do not account for such phenomena as the allosteric changes in overall shape of protein molecules, which contain thousands of atoms. When this allosteric behaviour was discovered, the change in our knowledge is better described, I think, in 'delving downwards' terms, by saying we have discovered something we didn't know before about the properties of atoms as they are exhibited in molecular groupings, than by the 'building up' statement that the atoms remain the same but we add something new and disconnected to them.

This reference to the importance of the overall shape of complex molecules, must serve as an introduction to another essential point in my argument. Recently, even the 'builders up', who accept the physico-chemical entities as a basis of all scientific knowledge, have realised that something more may be involved in them than the properties of mass, energy, etc., attributed to them in classical theory. This further component might be referred to as 'specificity' of spatio-temporal configuration. In the last twenty years or so, mathematicians and engineers have attempted to replace the rather undefined term specificity, which had been much used by biologists earlier, with a more precisely defined notion of 'Information'. Unfortunately, in order to achieve a precise definition capable of being utilised in a mathematical logical system, they have 'purified' the notion until it has become almost useless in connections with biology, or indeed in almost all contexts except that of messages; which was the main business of the Bell Telephone laboratories in which the originator of the theory, Claude Shannon, was employed. 'Information', as it emerged into the world of mathematics, is a measure of the degree of selection which has been employed in choosing some particular configuration out of a closed universe of possible configurations. It is concerned only with the specificity within a particular universe of possible specificities. For instance, the amount of 'Information' contained in the letter A is less if it is chosen out of the English alphabet of 26 characters than if it is chosen out of the Russian alphabet with 29. Moreover, the amount of 'Information', in this sense, has nothing whatever to do with bringing about any action outside the closed universe; that is to say, it has nothing to do with 'meaning', in any sense of that term. The information content of a message written in English words is just the specificity of the string of letters in which the words are spelt. Consider the two messages:

MEET HIGH MARKET TWELVE TEN
MEAT HIGH MARKET TWELVE TON

The differences in 'Information' are simply that the third letter from the beginning is an E in one and an A in the other, and the penultimate letter is E in one and an O in the second. 'Information' Theory has nothing whatever to say about the fact that the first is obviously about an appointment to meet at the corner of High Street and Market Street, and the second is a message from a wholesaler that the stocks are going off and had better be got rid of as quickly as possible.

This limitation in the meaning of 'Information' made it possible to develop a mathematical theory which is very useful in connection with transmission of messages along channels, but effectively ruined it as a word which is useful to apply in wider contexts. Rather unfortunately, the mathematical theory assigned to the measure of 'quantity of Information', a formula which was identical in algebraic form with one of the most famous formulae of thermodynamics, namely that for entropy. This at first led Shannon to identify the amount of information given out by a source with its entropy. Later, Warren Weaver developed an alternative interpretation, that the quantity of information contained in a message is the negative of its entropy. It was Weaver's rather than Shannon's interpretation which became fashionable, and the new word 'negentropy' was invented to mean 'quantity of information or negative entropy'.

The relevance of all this is that there is no doubt that reactions in living systems are very much concerned with the specificity rather than the mass or energy of the components. It is the specific arrangement of nucleotides along the chain of DNA which determines what that gene will do; it is the specific shape in three dimensions of a protein molecule which determines what sort of enzyme activity it will exhibit; and there are many other examples. For a time it became fashionable to discuss this sort of specificity in terms of negentropy, and some of the most penetrating minds, when they turned from physics to biology, were deceived for a time. Thus Schroedinger, in his elegant essay *What is Life?* in 1944, indulged in aphorisms such as 'life feeds on negentropy'. However, he soon came to realise that this is an inadequate way of looking at the situation, and he withdrew or at least greatly qualified the remark in the later editions of his book.

The main point is that the specificity with which biology is so deeply concerned is not a static specificity, with no meaning outside itself. It is rather the possibility of bringing about, or tending to bring about, a certain type of activity in appropriate things which react with it. It is, in fact, a specificity of instruction, the imparting of one particular program, or algorithm. We are returning here to a theme which has come up often in these lectures. Christopher, for instance, has been insisting that language is basically to do with programs or instructions, rather than with imparting descriptions from which nothing follows.

Of course the word information, as it is used in ordinary speech, often has some implication that the information will be useful as a guide to action. But it is pretty ambiguous in this context. In fact, during World War II, there was a useful distinction made in the slang of the RAF, which distinguished the 'Info', a lot of boring rigmarole about useless facts, from the 'Gen', the real stuff you needed to know to tell you how to operate. When we say that biological systems work by means of the programs or instructions incorporated in their components, this is a long-winded way of saying that it's the gen, not the info, that matters for them. It is not negentropy they feed on, but it might have made some sort of sense to call it gentropy, if I may coin an unnecessary word.

It is not only biological systems that feed on gen. There are some physico-chemical systems, which no one would dream of calling living, which very clearly do so too (possibly they all do, but I will not pursue this point here). Consider the min-erals making up that, at first sight, boring material, clay. They have been discussed in some detail from this point of view by Cairns Smith in his book *The Life Puzzle*. Clay minerals consist of crystals in which atoms of silicon, oxygen and a number of metals, such as aluminium, iron and various rarer and less frequent ones, are arranged in a three-dimensional lattice. The lattice is such that at any given time in the growth of the crystal its boundary is a flat two-dimensional plane, with a particular arrangement of these atoms at certain points on it. Now the forces at work are not terribly choosy about which particular atom goes into which place. At one particular point on the surface there might be an atom of aluminium or alternatively there might be an atom of iron, or some other substance. 'Ha!', the information theorists will say, 'this surface can encode a great deal of "information" '. So it can, but the point is that this is not mere info, it is gen. If there is iron instead of aluminium at point X, and the crystal is in a solution which allows it to grow by the deposition of a new layer of atoms on top of the old one, it is much more likely that another iron atom will take this place in the lattice of the next layer. The presence of iron at X is an instruction for building the next layer.

Whatever we imagine the first living systems to have been like, they must have been even more deeply involved in a traffic of instructions. Any type of hereditary material, be it DNA or anything else, which can be transmitted from one ancestral system to two or more daughter systems, must in effect contain instructions for its own copying. Moreover, in all the living things as they are on this earth, the copying system is carried out by mechanisms, such as enzymes, which operate by means of instructions built into them. Finally, systems which we consider worthy candidates to be granted the name living, differ from things like clay minerals in that they contain instructions, not only for copying, but for the elaboration of structures which can actively operate on surrounding materials. These new embodiments are what geneticists speak of as the phenotype. The crucial role of instruction-generated phenotypes as a fundamental aspect of living systems has been a dominant theme in recent discussions of the theory of general biology (see the four volumes entitled *Towards a Theoretical Biology*, edited Waddington, Edinburgh University Press).

The early stages in the evolution of life therefore involve not only physico-chemical mass, energy, atoms, and so on, but also specific instructions. We find the firmest evidence of mind when we look at the other end of evolution, as in our 'occasions of experience', and we are again, of course, fundamentally involved in a traffic of instructions. A knower does not merely sit down before the known, and observe it without comment or response. On the contrary, he brings to it certain predispositions, or interests, and observes certain characteristics more than others. The content he finds in the occasion demands a response. As Popper has put it, the 'prior knowledge' with which he comes to the occasion is such that what he receives from it is not mere information but instructions or challenges.

In the light of this discussion, the evolution of mind

79

appears as a transition from the instructional traffic involved in the very simplest living things, or even in the pre-biotic systems such as clays, to the much more complex traffic of instructions involved in our own occasions of experience. We can see two ends of the evolutionary range in similar terms. We have evaded the dilemma of considering the beginning of the evolutionary process as depending on nothing but atoms, forces and physico-chemical factors, while the other end involves something of a totally different character we call mind. One recent author who has advanced a similar view is Stephen Black. In his book *The Nature of Life,* he also draws attention to the importance of instructional traffic in all the processes of life (unfortunately he has not escaped from the fashionable convention of speaking of information when what he really means is instructions). His next step, however, is to expand the use of the word 'mind' to cover the whole range of situations involving instructional traffic from the very simplest to the most complex. This is hardly satisfactory, since, as we have seen, the simplest such situations occur in things like clay minerals, and it is hardly illuminating to speak of them having minds. When God fashioned us out of clay, he may have picked the right material to start from, but there was still a lot to do. What we need to do is to consider the nature of the evolutionary processes which have led from the simpler situations to the more complex ones.

In my second lecture last year, I sketched some of the broader characteristics of evolution. I pointed out that it is a process which occurs not in individuals but in populations of individuals. Any population contains a large variety of different hereditary potentials. Each individual as it grows up will develop some characteristics – making up its phenotype – which depend partly on its hereditary endowment and partly on the particular circumstances it happens to meet during its lifetime. And it is its phenotype which enables it to meet more or less successfully the demands made on it by its way of life, to leave more or fewer offspring, and to contribute more or less to later generations. The demands made on it depend on the way it spends its life. I gave as an example, a horse which can escape from its enemies by running away or by standing and fighting back. Which of these two alternatives it follows is a characteristic of its phenotype, developed gradually as the horse grows up, depending partly on hereditary predispositions and partly on external circumstances, such as other horses whose behaviour it copies, and so on. We describe these alternative forms of behaviour by saying that the horse opts for one or other of a number of goals – either the goal of winning the race by running away, or the goal of beating the predator in a stand-up fight. I argued that any set of behavioural acts in a living creature would only tempt us to ascribe mind to that animal, if they tend to bring about the achievement of some definable goal. If they were a mere higgledy-piggledy collection of responses to any instructions the surroundings happened to throw out, with no central theme or guiding principle, they would seem as mindless as the movements of a feather blowing about hither and thither on a gusty day.

In discussing the evolution of mind, we therefore have to consider first how goals arise. But then I shall want to raise the question whether having only one goal would be an adequate claim to having a mind.

As to the origin of goals, what we have to explain is the setting up of a system in which a lot of different and essentially distinctive instructions are woven together to a course of action which tends towards a definite end-point, or at least in a definite direction. How much of a problem is it to account for this? There have recently been two studies which suggest that some degree of coherence of disparate instructions into a degree of orderly behaviour, some degree of 'canalisation' of the resultant activity into a fairly definite direction, may occur spontaneously.

Stuart Kauffman set up a system which one can think of as a large number of light bulbs, controlled by on-off switches. The switches were connected by wires in such a way that there were two wires coming inwards to each switch, and the connections were made quite at random. It was also decided, by a random process such as throwing a dice, whether a light would be switched on when the lights from which it received signals were both on, or when they were both off, or when one was on and the other off, and so on. This system seems about as random and chaotic as it can be made. Kauffman tried the effect of turning on a random set of lights, then letting the whole system develop according to the rules built into it. He was astonished to find that in a surprisingly large number of cases, though not in all, the system very soon got into an orderly state, in which it went through a cycle, with one set of lights on, then changing to another, and to another and another and so on, but eventually coming back to the first set, after which the process was repeated. This is quite a high degree of orderliness, coming unexpectedly from a random set-up.

The other example which suggests that some degree of order may appear spontaneously is a theoretical calculation by Stuart Newman. He showed that in a system containing a large number of enzymes, with the particular characteristics that biological enzymes normally have, and operating under the sort of conditions we find in living things (in particular with low concentrations of their products), then whatever the conditions from which the system starts, it would soon follow one or another of a very small number of alternative paths of overall change.

These suggestions about the spontaneous origin of order

81

from chaos — spontaneous formation of goals — are still very recent and little understood. I think they are interesting in that they suggest that during evolution natural selection does not have to do the whole job of producing orderly goals of behaviour from a disorganised collection of immediate responses to haphazard instructions. However, even if natural selection is presented with a certain degree of spontaneous order as a basis, there is no doubt that well designed goals, such as the running-away behaviour of horses, have been moulded and perfected chiefly by the fact that individual horses that have behaved in this way have succeeded in leaving more offspring, thus passing on their propensities to later generations.

But if an animal always behaves in accordance with one definite unalterable goal, how much of a mind would we be inclined to attribute to it? Surely we wouldn't think it was being very clever. In fact, we might be tempted to say it was indulging in 'mindless repetition'. We would be much more tempted to think the animal had a worthwhile mind if it had at least two goals, and followed one or the other in appropriate circumstances.

Let us return to the horse which may run away or stand up and fight. We might discuss the situation in terms of a visual model, similar to the one I mentioned in my first lecture when I was discussing embryonic development, and went on from that to the problem of free-will. I suggested that one could consider the development of the different organs in the body, such as the brain, heart, muscles and so on, in terms of a landscape containing a branching set of valleys, which I called chreods, separated by watersheds. We can use the same sort of model to consider the development of different types of behaviour, directed towards different goals. A horse as it grows up has, let us say, the two possibilities of running away or standing and fighting. Horses at their present state of evolution are born with a tendency towards the running-away type of behaviour, and as they get older and more experienced in this mode of activity it probably gets built into them more firmly; it becomes more strongly chreodic, or more and more difficult for them to fall into the other behaviour and stand and fight. But clearly they would be showing more mental ability — and it would also be better for them from the point of view of natural selection — if they could run away from those animals they could be certain of outdistancing, but also adopt a stand-up-and-fight tactic to an animal that is faster than them and which they could not escape.

Thus the evolution of the mind must involve not only the formation of the goals, but also the development of alternative goals, and the ability to pick the appropriate goal under particular circumstances. In terms of our landscape model, natural selection will be attempting to produce a set of valleys which are definite enough, but which are separated only by low, rather than high and steep sided, watersheds. In fact it will be advantageous to the animal if the normal state of its central nervous system – its mental condition – corresponds to a point not at the bottom of one of the valleys, but some way up the hillside though perhaps not on the crest of the watershed. Then it can relatively easily move down into the most appropriate channel of behaviour to meet the particular circumstances with which it is confronted. In my first lecture I described the characteristic of the situations in which we feel we have to exert free will as being poised on a watershed between two or three alternative courses of action. The argument I have given here shows how the forces of evolution, operating through natural selection, will bring the mental apparatus of animals into such situations; that is to say, free will appears as a natural product of biological evolution.

Any nervous system which could function as a primitive mind in lower animals must have had certain characteristics if it were to be capable of evolving into the higher types of mind we have just considered. I shall have to skim over these very rapidly if I am to get on to another point which I want to make. But we can say that, at the most primitive level, it must have had a perceptual apparatus which incorporated a 'prior knowledge' which enable it to recognise single instructions from its surroundings, and to react appropriately to them; for instance, biting at and eating something recognised as food. As a first step in combining a number of different instruction-responses into a chreodic path to a more complex goal, it could perhaps get some assistance from the sort of order-out-of-chaos which Stuart Kauffman and Stuart Newman have been exploring. But surely it could not get very far in this direction without having some system of storing up, in a memory of some kind, the various activities it could do, and being able to pull them out again, and combine them with each other in all sorts of different ways, to see how they fit together. Do actions in some combination or sequence dovetail into each other so as to achieve a worthwhile goal – worthwhile for natural selection? The problem of mind – being intelligent – at this level is not only to find new ways of attaining already accepted goals, but puts a premium on the still greater flexibility of discovering new goals.

I will tell the tale of the evolutionary origin of the birds; or rather, one of the more plausible tales, because I don't think the experts have quite decided exactly how it did happen. But one of the ways it may have happened concerns a group of little reptiles, rather like lizards, which had larger hind legs than forelegs, and which normally ran about on these back legs. Suppose they started using their forelegs to work up speed when they were running away from a nasty bigger reptile who was trying to catch them. And suppose the scales on the arms grew longer, into something a bit like feathers, to help them get the benefit of beating the air effectively to push them along. And natural selection pushed this development further, until, one day, some of them found themselves taking off, and becoming airborne. It must have been very disconcerting; they probably ran the risk of crashing in considerable disorder, and getting gobbled up. But a really clever little lizard, full of mind, must have said: 'Hey, we've got something here', and set about finding how to fly. To attain this brand-new goal, he may have had to change quite a lot of his previous routines; for instance, beating his arm-wings in unison instead of one after the other in time with his legs. In order that such an evolution could be possible, his mind had to be able to do two things. It had to be able to re-organise itself around a new sub-goal, to fly, within its old main goal, to escape; and it had to be able to re-arrange its detailed activities so as to achieve this new sub-goal, to change the timing of its arm movements, for example.

Now evolution by natural selection does not reward *only* flexibility. Certain shellfish which live in deep mud have been able to go on successfully living in deep mud unchanged since the Cambrian era till today. But evolution does also reward flexibility, when it comes off. Our clever lizard started the whole kingdom of birds, one of the main types of higher animals. It is in this context, I think, that we have to see the origin of human language, which was the point I wanted to get on to, because with the evolution of language, the whole nature of the best minds to be found in the living world is profoundly altered.

I shall not discuss the evolutionary origins of language in any detail, because I think Tony Kenny will be talking about it tomorrow. I just want to make, quite briefly, one or two points. In the first place, language is obviously a very powerful instrument in the enterprises of re-organising and re-structuring of older ideas and activities which are necessary for flexibility in forming and attaining goals. The use of symbols, such as words, is a very convenient means of access to a memory store; and the arrangement of words into syntactical sentences is a good way of examining the structural relations between those items. To get an idea of the power of the method, compare the rate of improvement of a human skill which can be handled with symbols, such as the ability to solve mathematical equations, with that of one which cannot, such as, say, high jumping. Both do improve as time passes, but the first, which can be taught by words, improves enormously more rapidly than the second.

The appearance of symbolically transmitted culture brought about a speeding up, by orders of magnitude, of the processes of evolution in the species which enjoyed it. I have discussed this in several books since Julian Huxley, and I first

84

suggested some thirty years ago that language amounts to a new and faster genetic system. I shall not labour the point here, but I want to remark that the advantages of cultural transition were not obtained by mankind without a price being paid. When the audience was asked to submit questions at our open discussion, someone referred to Arthur Koestler's suggestion that the recent evolution of man's brain incorporates a basic anatomical flaw, which is the cause of the many ills which have persistently beset civilisations based on transmitted culture. I think that it is not necessary to look any further, to find the root of these troubles, than to the fact that symbolic transmission is an evolutionary novelty, and that man is still trying to master this new skill, and is floundering about in a thoroughly clumsy way, like the first bird-lizards who found themselves unexpectedly air-borne.

The main difficulty in becoming a skilful performer, who never loses control and crashes, arises I think from the method mankind uses for turning a physical occurrence, such as a noise, into a symbol, a word which has a meaning. In practice this essential conversion depends on associating the noise, in the mind of the very young baby who is being initiated into the world of culture, with some sort of authority which controls its acts, such as a parent. I have argued in detail in my book *The Ethical Animal* that it is from this process — noise becomes word by association with personal authority — that man derives his characteristically human notions of a personal God, and of self-transcending criteria of good and evil. But the process, at this stage of man's evolution, very often does not work smoothly. It may give rise to a Freudian 'super-ego', and all sorts of irrational obsessions, fears and hatreds. There is no reason to expect that keeping aloft in the realms of spirit is any easier for man than flying through the air was for lizards. Man has already taken some nasty tumbles, and he may finally crash fatally; but on the other hand, our descendants may attain as much mastery over the world of good and evil, belief and freedom, as did the lizard's descendants, the sea-gull and the kestrel, over the thin and dangerous air.

One final point. This lecture was announced with the title: 'Does Evolution have a Goal?' When I came to write it, it did not turn out quite like that. I had discussed that before, in *The Ethical Animal,* and I wanted today to say something different. But I should like to finish by asking whether there is anything inevitable, or foreseeable, about the appearance in evolution of the human mind, which is so much a product of the cultural mode of socio-genetic transmission.

The cybernetic mechanism of evolution which I have sketched earlier, in which phenotypes form goals, and populations are put through the hoops of natural selection to find those members which have goals which they are able to achieve, would be expected to produce all sorts of different life-styles and ways of succeeding in them. One could regard the whole kingdom of living things as a series of feelers, extending in all directions to explore the realm of possible modes of staying alive and reproducing. Now I think one can say, with hindsight, that it is obvious that a cultural-symbolic mode of transmitting instructions from one generation to the next is not only a possible mode, but one which would provide much greater flexibility in fewer lifetimes than the mode employing the DNA-RNA-protein machinery which most of the living world relies on. It follows that, if the realm of possible life-styles is being explored all over, something will eventually stumble on this possibility.

But I see no reason why the cultural-symbolic mode should have taken just precisely the form which it actually has taken in the evolution of man. The essential point in it is that something or other, which is transmittable from one individual to another faster and more frequently than DNA, should become endowed with the power of being a symbol, which conveys instructions. Noises, or visual signs like letters or ideograms, are good candidates for suitable carriers of symbolic meaning. But I am not sure whether there is any compelling *a priori* reason why the conversion of these neutral physical stimuli into symbols should have been by associating them with personal power or authority. Could there have been an evolution of another type of language, in which a noise became a word, not from its association with a parent who controls one's behaviour, but, perhaps, by being associated with the definiteness of different objects of perception, which the baby is also discovering at about the time it learns to talk? The noise 'cup' would get that extra something which turns it into a symbol, not by derivation from the authority of a parent who was trying to make you drink out of it, but by association with the realisation, which must be extremely powerful and impressive when it first dawns on the awakening mind, that certain groupings of sensory inputs add up to something which you can pick up as a separate entity, which can contain fluids, and so on. If symbols and language had evolved in this way, rather than in the way they have actually evolved, our minds would have a basic tendency towards polytheism, seeing God in everything, and a non-personal, non-controlling God at that, instead of towards the monotheistic Law Giver who has such a hold on the Man who has actually appeared on the stage of evolutionary history.

Perhaps, indeed, something of this second course has entered into the evolution of language as we know it, and slightly, though I would say not sufficiently, blunted the cutting edge of the dangerous weapon of authority-based symbolisation.

Discussion

LUCAS

I think the points which will be most interesting to us, who are for the most part not biologists, in the consideration which Wad, himself an eminent biologist, has given to the nature of the evolution of the mind, are going to be the points he brings up to avoid being forced into a reductionist position. He is at pains to say that evolution is to be accounted for not only in terms of Darwinian theory but also in terms of information theory; and then he rapidly goes on to refine this and say that information theory isn't the stuff that goes on in Bell Telephone laboratories, it is not negentropy but rather gentropy, throwing away most of the stuff we are told, as being mere info, and laying all emphasis on just the know-how of the important things to do. Here, I think, in the rejection of the sort of rather pale academic information and the emphasis on the real full-blooded passions of the mind, we might once again discern at work in Wad's mind, the influence of David Hume. This move towards the imperative rather than the indicative mood which is stressed by Christopher is of very great concern to us, as giving us some idea what the mind is, and also the questions we need to ask in coming to terms with the facts we know about our ancestors and their pre-history.

The key notion which we haven't yet illustrated enough is that of goals. Wad wants to say it is just not simply an instruction — after all, those rather boring clays are able to carry out certain self-replication instructions and this he is prepared to reject, if it's offered as a typical phenomenon of mind, as being merely mindless repetition. Rather than this, we have got to have a certain choice between alternatives; and he is very much concerned to argue that there should be some element of freedom in this choice — both in the account of the early ancestors of the horse, and in the account of the way the birds came. In each case, he quite deliberately uses very personal language.

This point I want only to allude to, because there is not time to extract any philosophical meat out of it; but it is very significant in our attempt to make out what lessons we should learn from evolution, what are the questions that we ought to ask. What we are being shown is that there is some element of free choice which characterises even rather primitive living organisms, but more pre-eminently those that we are prepared to ascribe minds to. These are subjected in the first and classical case to the discipline of natural selection, and then, with man and more or less with man alone, to the further form of information control which is constituted by the use of language. Like Wad, I shall not trespass on this because it is Tony's preserve, and it would be a pity to broach the subject before him; but I just want to raise one point where I think I can quarrel with Wad without stealing any of Tony's clothes.

We have a new version today of the origin of sin. No longer is it either that Adam needed apples or that Prometheus stole the fire from the gods. It is not even the normal fashionable Freudian account that one can read week by week in the Sunday papers, which is something to do with 'potty' training and things like that. Nevertheless, like that, it goes back to one's very early years, but it is in the instruction in the use of language that all our troubles are said to arise. It is authoritative-based symbolisation which gives rise to our self-transcending criteria of good and evil. These, although a powerful force for good, have somehow come adrift and made us subject to quite disastrous lapses of judgement and temper. I don't think this is an adequate account for a number of different reasons; I think there is a false antithesis between the transcendent concept of right and wrong which, I agree, the symbolic nature of language enables us to handle, and what Wad would rather have us adopt, some sort of self-generated concept of right and wrong. (If we are going to have the word 'transcendent' I suppose we should also have the word 'immanent'.) These terms are not opposed. Although the use of language is something which does enable us to form the concept of right and wrong, it is not either essentially, or even as a matter of contingent fact, instilled simply by authority. Some authority no doubt, but also very largely an element of exploration. After all, no amount of authority will teach some people to speak; and the schoolmaster can use his rod but fail to instil anything into a great many impenetrably thick intellects which, if nothing else, stand as counter-examples to Wad's thesis.

WADDINGTON

I am really surprised that John will not allow me to associate the idea of sin with the acquisition of language. I thought I was simply rephrasing the Book of Genesis. After all, the Fall of Man and the expulsion from the Garden of Eden followed eating the fruits of the Tree of Knowledge; and if that isn't acquiring language, I don't know what is.

Now, to go back to John's point about goals and personal choice. I was quite deliberate in using personal language in this connection, as he said I did. I think I have argued at one of these lectures — I have certainly argued it in other places — that even in the apparently simple act of perception, something is involved that is very similar to purpose. The recognition of an item in perception involves accepting that what you are perceiving is near enough to some internal model, which has a chreodic character, so that it can act as a 'centre of attraction' into which anything that falls within a certain neighbouring area of variation gets sucked in. This bringing into line of nearby variants is something very like the behaviour of a system which has a goal. So if you believe, as I do, that science is based on experiments, on experiences involving perception, not on hypothetical constructs such as atoms, electrons and so on, then I do not think one need be frightened to introduce ideas like personal choice, or at any rate ideas like goals, right at the basis of one's whole concept of scientific understanding, because such ideas come into the very basic notion of perception.

LONGUET-HIGGINS

May I raise another point, of a rather different kind? You touched on the question whether pre-biotic evolution, or should one say the inorganic precursor of evolution, would inevitably or most likely produce life. This is a question which cannot possibly be answered without reference to the prevailing conditions and how long you are prepared to wait. There is, for instance, the 'nitty-gritty' question of whether there's any water about. It looks as if the sun, for instance, is a great deal too hot to permit the evolution of life, even over a very long time. The moon is too cold: maybe there was life there once but we don't think the conditions are right any longer. These are issues which one can only discuss, it seems to me, in coldly physico-chemical terms; I doubt if anyone is going to arrive at the answer by purely information-theoretical reasoning, without reference to the way things were. That's why I felt a little uneasy when you dismissed the physics and chemistry as fallacious reductionism. Information theory, which I agree we now have to use, wouldn't be relevant until there was something about which to information-theorise, a self-replicating system of some sort, and no-one has yet accounted for the origin of such systems, as far as I know.

KENNY

Like John, I welcomed it very much when Wad drew our attention to the ambiguities of the word 'information' and showed the dangers of applying Shannon's information theory to the sort of things we are talking about when we use the word 'information' in ordinary language. But I thought that Wad then fell into the same error when he introduced his new word 'instruction'. He was using 'instruction' in such a broad sense that while it is a case of instruction if I get into a taxi and tell the taxi driver to drive to the Carlton Hotel, it is equally a case of instruction if I get into my own car and shift the gear lever and turn the steering wheel. If we are talking about the evolution of mind, we're surely talking about what makes the difference between taxi drivers and taxis and, with all respect, I thought there wasn't a word in Wad's paper to explain the evolution of mind.

There can be at least two views about what mind is. In these discussions there have been those of us who wanted to identify mind with consciousness and those who wanted to identify mind with such things as the ability to use language. In either case, if one has to explain the evolution of mind one must show how beings which clearly did not have minds (which clearly did not have consciousness, for example, or which clearly did not have language), could turn by processes of natural selection into populations which did have minds (which did have consciousness or which did have language). I didn't hear any point in Wad's paper where he took such a case and showed how it could come about. I thought he was going to do this when he began talking about the evolution of birds, but instead of using the evolution of birds to explain the evolution of mind, he used, as John pointed out, the presence of mind to explain the evolution of birds.

WADDINGTON

I suppose that eventually we shall have to tackle this problem of trying to define the mind. Tony has said that there are some who want to define it as the use of language, and others want to define it in terms of consciousness. Now I should not want to define it as either of these. Certainly the use of language is something which certain types of minds can do, but to me that does not mean that all types can. Consciousness is something I know very little about. Whether the first 'clever little lizard' was conscious or not I can only guess; possibly it is easier to suppose that he was. But honestly it seems to me that we really don't know, and I don't see how we ever could know.

I am not prepared, as yet at least, to give a definition of mind. I tend to see it as in some way connected with the integration of a lot of intentions or goals, or with the integration of a lot of responses to instructions, into complex systems which lead to some definite end-point. Or rather, I'd say that any but the most absolutely primitive mind has got to be able to integrate responses so that they can lead not just to one, but to several different end-points; it must really have alternative goals. Now, whether they are conscious goals or not seems to me a question you cannot usefully ask, because there is no way of getting an answer to it, except about oneself, and about people who can talk to one (supposing one believes what they say); while it seems to be clear that there are many animals which can perform the sort of integration of responses I refer to but which do not use language. When I said something about the origin of language, I was not, in my opinion, discussing the origin of mind, but only one stage in the evolution of mind. However, Tony will be coming back to the question of language, and when I reply to his lecture we shall be able to return to this theme.

ANTHONY KENNY

Seventh Lecture. The Origin of Language

In the first four lectures, we discussed the development of mind in the individual human being. In this final batch of lectures, we are considering the origin and future of the mind in the species.

Waddington attaches great importance to the notion of goals as defining mentality, and therefore when he was talking about the origin of mind he spoke in particular about the possible origins of goal-directed behaviour. In my first lecture, I offered a rather different definition of mind — the mind, I said, is the capacity to acquire intellectual abilities, and by intellectual abilities I mean activities involving operations with symbols. So now that it's my turn to talk about the origin of mind, what I have to discuss is how people can have begun to operate with symbols; I have to face the problem of origin of language.

No-one really *knows* anything about the origin of language. John Lucas reminded us that, in the book of Genesis, we're told that God fashioned all the wild beasts and all the birds of heaven and brought them to Adam to see what Adam would call them. Each of them was to bear the name that Adam would give it. I don't know if John takes this to be the literal truth about the matter. I imagine that probably few people do, but we haven't yet found an account to replace the Genesis myth.

'The biological history of language cannot be revealed through a random comparison with animal communication', says a biologist. 'Reconstruction of the origin of language is impossible except for some very simple determinations. This is because of the following limitations: (1) the size and shape of the brain furnish no secure clue about the capacity for language; (2) given morphological peculiarities of the CNS do not bear a fixed relationship to behaviour; (3) even if we had direct knowledge of social structure or cultural complexity of the societies of various fossil men, we could not draw conclusions about language as we know it today. Different types of communication might have prevailed at those times.' (Lenneberg, *Biological Foundations of Language*, p. 266.)

The linguists are similarly pessimistic. Lyons has written recently: 'What one might call the "official" or "orthodox" professional attitude of linguists to evolutionary theories of the origin of language tends ... to be one of agnosticism. Psychologists, biologists, ethologists and others might say, if they so wish, that language *must* have evolved from non-verbal

91

communication; the fact remains that there is no actual evidence from language to support this belief.' (Hinde, *Non-verbal Communication,* 76.)

Obviously, I'm not going to claim to know where language came from when the experts can only guess. On the contrary, I'm going to suggest this evening that if anything Lenneberg and Lyons, in the passages I've quoted, underestimate the difficulty of explaining the origin of language by natural selection from animal communication systems. It isn't just that languages themselves don't evolve in the appropriate way, or that we don't know much about the brains of primitive men because brains don't fossilise. The difficulties are difficulties of principle: language as we know it has features which are *prima facie* inexplicable by natural selection.

I complained yesterday that Waddington didn't explain how language might have originated among a community that didn't have language. He did talk about the acquisition of language by children: but as his account presupposed that they learnt language from their parents, it offered no explanation of the origin of language in the population as a whole. However, in his first lecture he did make such a suggestion. He suggested that language might have developed by 'the formation of a chreod surpassing the needs of integration which brought it into being'; as a floating animal with three stiff-jointed rods would greatly increase its stability if they formed a triangle, perhaps beyond anything actually required for protection.

'It is not impossible', Waddington went on to say, 'that the universal generative grammar spoken of by Chomsky originated in a rather similar way; in some set of rules, which were at first favoured by natural selection because they made it possible to elaborate slightly on the degree of communication that could be carried out by ungrammatical grunts, shouts and so on, but which then turned out to be enormously more powerful than required by the immediate needs of the situation.'

As I remarked at the time, I think that in this passage Waddington somewhat misrepresents the nature of Chomsky's theory. He seems to treat universal grammar as differing from particular grammar rather as Esperanto differs from English. Now there are, as we shall see, difficulties in the idea that a grammar, like that of English or of Esperanto, might have evolved. But the difficulties in thinking that a universal grammar might have evolved are difficulties of a very special kind. I should like to insist once more on the highly abstract nature of the structure postulated by Chomsky. A child has to acquire the grammar of a language: he has to internalise the rules governing its use. To do so, Chomsky says, he must unconsciously frame a hypothesis about the nature and structure of the linguistic data fed him by his parents and other humans in the environment. Now the data available to him are too fragmentary, according to Chomsky, for him to be able to acquire the grammar unless he had some innate knowledge of the type of hypothesis that is acceptable. Universal grammar, then, is itself not so much a grammar as a sieve through which candidate hypothetical grammars must pass. To see the relation between particular grammars and universal grammar, imagine that a Ministry of Sport sets up regulations governing the forming of rules of particular games. Particular grammars will then be related to universal grammar as particular rules are related to Ministry regulations. Neither the Ministry's law nor the set of provisions it will contain will be sufficient to enable one to play any game; the sort of provisions it will include will be 'no rule of any game may be longer than fourteen words', 'no infringement of a rule of any game shall be punishable by death', etc.

Whatever difficulties there are about explaining by natural selection the internalisation of the rules of a particular language, it is surely doubly difficult to explain in that way the internalisation of a set of rules about rules. For the biological utility of having an internal sieve to assist in the mastery of grammars cannot antedate the existence of grammars themselves. To try to explain the origin of universal grammar by natural selection seems as least as difficult as trying to explain the utility of a light-sensitive cell to an animal living in an environment totally without light.

However, Chomsky's postulate of universal grammar is still a controversial hypothesis. If the innateness hypothesis is incompatible with explaining the origin of human language by natural selection, one may be inclined to say: 'So much the worse for Chomsky's hypothesis'. So I pass to the second, and major objection that I wish to raise against evolutionary explanation of the origin of language. This is based not upon a dubitable hypothesis but upon a truism: the truism that language is a social, conventional, rule-governed activity. The importance of this truism does not seem to have yet been fully appreciated by all biologists and linguists, although it has, of course, been much stressed by philosophers in recent decades. Three philosophical works are worth mentioning in illustration: Wittgenstein's *Philosophical Grammar* (1933), Bennett's *Rationality* (1964), and Searle's *Speech Acts* (1967).

The most important difference between language and other media of communication is that language involves the use of symbols. Symbols differ from other signs (traces, clues, symptoms) in being conventional and not natural signs. When we say that language is conventional we do not mean that it was set up by some primeval linguistic contract on the style of Rousseau's social contract. Such a suggestion would be grotesque, not to say self-contradictory: obviously one cannot make a contract that a certain word is to mean X unless one already has the linguistic means of referring to X. Rather, when we say that language is conventional we mean that the behaviour of language users is rule-governed. This means, among other things, that language-behaviour is not as such governed by causal regularities: one can only keep a rule if one can also break it. There is no law of nature to the effect that our utterances have the meaning that they have.

There are several differences between rules governing behaviour and natural laws governing phenomena. One is that it is possible, and it often happens, that rules are violated. Short of a miracle it isn't possible for a natural law to be violated. If you find an apparent violation of a natural law you have an indication that the law has been wrongly stated. The occurrence of a violation of a rule, on the contrary, is no evidence that the rule has been wrongly framed.

Another difference is that if someone's behaviour is governed by rule he must be to some degree conscious of the rule. This does not mean that he must be able to formulate or enunciate the rule: it is notoriously hard even for a fluent language speaker to enunciate the phonological, syntactic and semantic rules he uses. What it means is that the user of a rule must be able to distinguish between correct and incorrect applications of the rule; he must know the difference between following and violating the rule if he can be said to be using the rule at all. In contrast, one can be operated upon by a natural law without having any consciousness of it at all. For it to be true that mammals are procreated as a consequence of sexual intercourse there is no need for mammals to have any knowledge of this fact; on the other hand if a community operates a certain transformation rule for interrogatives they must be capable of distinguishing between well-formed and ill-formed interrogatives.

Now the rule-governed nature of languages makes it difficult to explain the origin of language by natural selection. I shall say why in a moment: let me point out now that in my account of the nature of convention, I have not said anything about innateness. I contrasted conventional rules with natural laws, but I am not using 'conventional' to contrast with 'innate'. Something might be innate in the sense of not being acquired by learning, and yet be conventional in the sense that it was an activity in accordance with rules and not a mechanical procedure.

The difficulty is this: the explanation by natural selection of the origin of a feature in a population presupposes the occurrence of that feature in particular individuals of the population. Waddington has earlier given us examples about the development of the length of legs in horses. One can very easily understand how natural selection might favour a certain length of leg; if it were advantageous to have long legs, then the long-legged individuals in the population might outbreed the others. Clearly, where such explanation of the occurrence of features is most obviously apposite, it is perfectly possible to conceive the occurrence of the feature in single individuals. There is no problem about describing a single individual as long-legged, or as having legs *n* metres long. (There may be a problem about the origin of the individual long-legged specimen, a problem which may or may not be solved by talking of random mutation; what I am concerned with is not the origin, but the conceivability, of the favoured individual specimen.)

Now it does not seem at all plausible to suggest, in a precisely parallel way, that the human race may have begun to use language because the language-using individuals among the population were advantaged and so outbred the non-language-using individuals. This isn't simply because of the difficulty of seeing how spontaneous mutation could produce a language-using individual; it is the difficulty of seeing how anyone could be described as a language-using individual at all at a stage before there was a community of language-users. For consider; I am using language now in lecturing, and that I am doing so depends no doubt on decisions of my own and is conditioned on all kinds of ways by my own physiology; but whatever I did, whatever noises and gestures I made, they could not have the meaning my words now have were it not for the existence of conventions not of my making, and the activities of countless other users of the English language. If we reflect in this way on the social and conventional nature of language, then we begin to see something very odd about the idea that language may have evolved because of the advantages possessed by language-users over non-language-users. It seems as absurd as the idea that golf may have evolved because golf-players had an advantage over non-golf-players in the struggle for life, or that banks evolved because those people born with an innate cheque-writing ability were better off than those born without it.

We don't, of course, think of games like golf and institutions like banks as having been evolved; we think of them as having been invented. I don't wish to suggest that the origin of language can be explained in the same way. In order to be able to invent an instrument for a particular purpose you need to be able to conceive that purpose in advance and devise the invention as a means to it; whereas it doesn't seem that someone who didn't have a language could first of all conceive a purpose which language would serve, and then devise language as a means to serve it. Some human procedures were hit upon by accident, as in the legend that someone discovered the utility of roasting pork by having his house catch fire with his pig inside. It doesn't seem that language could originate in the same way, because such unexpected fortunate discoveries can surely occur only in the case of causal results, not of rule-governed activities like language. One can't conceive of a man's being the first accidentally to follow a set of linguistic

95

rules as one can conceive of him being the first accidentally to set fire to his house.

The difficulty that I posed about the natural selection of language may seem to be one not peculiar to the present case. If the difficulty is that an individual's behaviour isn't linguistic behaviour except in the context of the behaviour of others, is not this true of all kinds of social behaviour, including some of which dumb animals are undoubtedly capable? A courtship ritual, one might say, would not be what it is without the responsive behaviour of the mate. Nor, for that matter, would sexual organs be sexual organs were it not for the corresponding parts of the anatomy of the opposite sex. For all that, there seems no special difficulty in postulating the evolution of sexual organs or courtship rituals.

Both of these parallels, I think, break down: to see why will take us a step further. Sexual anatomy, like long legs, is conceivable and describable without reference to the counterpart anatomy, though of course its usefulness for procreation cannot be brought out without reference to the counterpart. The difficulty concerns utility, not conceivability. It is not the sharing of rules between language-speakers that would provide the linguistic parallel, but such things as that an animal could not use for species-specific communication sounds that were inaudible to other members of the species.

The parallel between language and ritualised animal activity is a closer one, though it too breaks down. Ritualised activities, in fact, occupy an intermediate position between language and causally efficacious behaviour. They appear often to be displacement activities: that is, they are causally inappropriate to produce the particular effect currently desired. In this rituals resemble language in not being part of a causal mechanism operated to produce an effect. But unlike language they do not seem to be rule-governed, that is to say, they don't seem to be accompanied by any behaviour expressive of a conscious discrimination of violation. Moreover, the ritual activities can be given precise behavioural specification (e.g. 'the two partners stand with their heads raised, the beak pointing downwards and the head turned away from the partner', J. Maynard Smith, *The Theory of Evolution*, p. 169, etc.). Linguistic activity cannot be behaviouristically described in this way, because very many different behaviours may be the same linguistic act. For instance, I use the expression 'No Smoking' whether I say it or write it, but the behaviour involved in writing it is totally different from that involved in saying it. This does not mean that it is the invention of writing, rather than the origin of speech, that marks the boundary between human language and animal communication. It means rather that the way in which human language is conventionalised is quite different from the way in which animal behaviours are ritualised. For part of what we mean when we say that human language is conventional is that *any kind of behaviour* will do provided that it obeys the right rules. Not only speaking and writing but morse code, semaphore, American sign-language, etc., can be used to utter the same sentence, given the appropriate conventions. Indeed, it underestimates the crucial function of rules in language to say that language is *governed* by rules. Language is rather *constituted* by the rules. Take any human sign language: you can change anything in the sign apart from the rules governing its use and the language will remain the same. Courtship rituals provide no parallel to this.

I conclude then that the difficulty in using the principles of natural selection to explain the origin of language is a special one, affecting language in a unique way by comparison with other phenomena natural selection is called on to explain. I do not say that it is an insuperable difficulty; but it is a difficulty which, so far as I know, has not yet been seriously faced.

The difficulty I have raised remains a difficulty whether one conceives language, in the individual, as a totally learnt phenomenon or as one involving an innate component. But as I have said the difficulty is compounded if one accepts Chomsky's view that the facts of language learning in the individual can only be explained if we postulate an innate mental structure, or universal grammar.

Christopher is fond of reminding us that in evolution ontogeny recapitulates phylogeny. In the case of the origin of language it seems that this cannot be the case. However far we go with Chomsky in postulating innate mechanisms we cannot deny that what children acquire, when they learn language, is the language of their parents: it is a language spoken in their environment. Whatever it was that the first language-users acquired, it was not a language taught them by their parents. Learning, of course, in general, does not demand teaching as a correlate: a rat may learn its way around a maze without a teacher by trial and error. But language could not originate by trial and error learning; because the notions of trial and error presuppose stable goals which successive attempts realise or fail to realise (the acquisition of the food pellet, etc.). But there is no independently specifiable goal to which language is a means: the communication of thoughts cannot be regarded as such a goal, because there are so many thoughts which can only be expressed in language. In particular, one can't have the goal of *acquiring a language* because one needs a language to have that wish in.

In 1967 Lenneberg remarked that it was a weakness of the theory that human language was continuous to animal systems of communication, that the examples cited in support of the theory were drawn from all over the animal kingdom, birds, insects, fishes, mammals, in complete disregard for phylogenetic proximity to man. It might be thought that his argument has been weakened since he wrote because of the successes of Washoe and Sarah: chimpanzees, reasonably close to man phylogenetically, who have shown remarkable abilities to mimic language. Now there is some dispute as to how far these two chimps have really mastered syntax; but let us waive this and accept for the moment that their linguistic skills are beyond reproach. The results still seem irrelevant to the prehistory of language for two reasons. The first, and less important, is that the modalities in which they have learnt language are very different from any known to any but very recent generations of humans, thereby making any phylogenetic continuity improbable. The second, more important one, follows from the point I have just been making: Sarah and Washoe were undoubtedly *taught* by the human families with whom they lived; and the demonstration of the ability to be taught a language in the ancestors of human beings would not go any way at all to explain the origin of language itself. For our problem was one that would arise even if there had been *men* around for generations before language started.

Striking as the successes of Sarah and Washoe have been, they do not in themselves have anything to tell us about the origin of language. The possibility of teaching human language to a chimpanzee no more proves that human language evolved from a pre-human communication system than the possibility of teaching a chimpanzee to ride a bicycle shows that bicycles evolved from pre-human transportation systems.

The difficulties in explaining the origin of language by natural selection are masked by the language we use to describe the phenomena. First, we tend to group together quite different human skills and performances; second, we tend to use abstract descriptions of human and animal communications in ways which conceal their differences.

An example of the first difficulty occurs in Popper's paper *Clouds and Clocks,* as Chomsky pointed out in *Language and Mind.* Popper had argued that the evolution of language passed through several stages including a lower stage in which vocal gestures were used for the expression of emotions, and a higher stage in which articulated sound was used for the communication of thought. Chomsky remarks: 'His discussion of stages of evolution of language suggests a kind of continuity, but in fact he establishes no relation between the lower and higher stages and does not suggest a mechanism whereby transition can take place from one stage to the next. In short he gives no argument to show that the stages belong to a single evolutionary process . . . In fact it is difficult to see what links these stages at all . . . There is no more of a basis for assuming an evolutionary development of "higher" from "lower" stages in this case than there is for assuming an evolutionary development from breathing to walking. The stages have no significant analogy, it appears, and seem to involve entirely different processes and principles.'

Another instance of the second difficulty occurs in the celebrated 'design features' isolated by Hockett for the comparative description of human and animal communication systems. In two papers (1960, 1968 with Altmann) Hockett isolated sixteen characteristic features of human language with a view to investigating which of them could be found also in animals. The sixteen are: (1) vocal-auditory channelling; (2) broadcast transmission and directional reception; (3) rapid fading; (4) interchangeability — speakers can be hearers and vice versa; (5) complete feed-back — the speaker hears what he says; (6) specialisation — the linguistic signals don't do any non-linguistic work; (7) semanticity; (8) arbitrariness — there is no resemblance between elements in language and their denotation; (9) discreteness — the signals are digital rather than analogue; (10) displacement — the signals refer to things remote in time and space; (11) openness — new messages are freely coined and easily understood; (12) tradition — the communication system can be passed on by teaching; (13) duality of patterning — combinations of meaningless elements can be meaningful; (14) prevarication — the ability to lie or mislead; (15) reflexiveness — the ability to talk about language; (16) transfer of learnability to other language-systems.

These features obviously provide a very useful framework for investigation. But in the light of what we have been saying, the most striking thing about them is that rule-governedness, *the* most characteristic feature of human language, is not mentioned at all. It is possible that rule-governedness is implicit in some of them, e.g. 'prevarication', 'displacement' and 'semanticity', but Hockett does not seem to be aware of this (he defines 'semanticity' in terms of associative ties).

It is possible that the use of the language of communication engineering in biology is a source also of the masking of difficulties. At an earlier meeting we discussed in what sense one could speak literally of a genetic *code*, for instance. The reason for calling it a code is that there is no known law specifying which group of bases in a strand of DNA determines which protein (string of amino acids). The correlation seems quite arbitrary. But that is not enough to make it a code in the sense of a *language*: for that to be so the operation of the DNA would have to be by means of rules (like a chain of command) and not by causal mechanisms (like that of a template). To see the difference think of a piano and a pianola. The holes in a pianola roll are correlated with the notes the pianola plays just as the notes on the page of the score correlated with the notes the concert pianist plays. But the holes in the pianola roll are not symbols in the way that the notes on the page are.

In general, the fact that X is isomorphic to and merely arbitrarily connected with Y does not make X mean Y or be a symbol for Y. In addition it is necessary that X should be linked with Y by convention, and that demands the intervention of voluntary agents acting intentionally according to rule.

If this were not so, the world would be full of symbols. Given suitable conventions, the fact that this glass is on the table might express the fact that I am sitting on that chair. That is, if the glass stands for me, and the table stands for the chair, and the relation of being on top of stands for itself, we might conceive the glass on the table as a false sentence saying that I am sitting in the chair. The situation has the right logical multiplicity to represent, and the connection between the two situations is entirely arbitrary. But of course that is not enough. The glass on the table isn't a sentence, and that is because we haven't set up the appropriate conventions: we don't use it as a symbol.

For something to be a linguistic representation of a state of affairs it not only needs to have the appropriate abstract structure, it needs to consist of elements conventionally correlated with elements of the structure to be represented. And conventions can only be set up by those who can *use* symbols: in the normal case, by human beings with the parts and passions of human beings. It is for this reason that I have denied, and deny, that the elements in a computer bearing particular information *mean* that information and are symbols for that information; and thus I have maintained that the output of a computer is only meaningful and symbolic to the extent that the designers and programmers of the computer have set up the appropriate conventions.

I conclude that there are two things which language must have: (1) an abstract structure capable of translation into diverse behavioural modalities; (2) a user who is capable of giving meaning to that abstract structure, by using it in rule-governed behaviour of the appropriate sort. Non-human animal systems of communication in the natural state appear to me to lack both of these features. If human language has evolved from non-human communication systems then at some time the activity of following rules must have been produced by natural selection. And there seem to be profound difficulties of principle in seeing how the practice of following rules could have originated in such a way.

Discussion

WADDINGTON

I agree with Tony that a great deal remains mysterious about the evolution of language. You have heard enough of my ideas already to know that I do not think that 'natural selection of random gene mutations' is a magic formula which clears up all the problems. But in connection with Tony's arguments, I first ought to remind you that the evolution of a great many other things besides language remains very mysterious too. We cannot give a full evolutionary explanation of the origins of any

100

of the major phyla. How did the worms originate? How did the insects, or the vertebrates? We do not really know in any detail. It seems that the evolution of these major changes of general organisation happens very rapidly; the system goes very quickly over from one strongly chreodic organisation into another, so rapidly indeed that very few intermediates were ever formed, and one hardly ever finds them in fossils. In a few places we have one or two intermediates, for instance between reptiles and birds, but as a general rule the intermediate steps in major changes in evolution are very little known.

There are some people, of course, who take this as definite evidence that evolution is not the whole story, and that there must have been Creation as well. In recent years there seems to have been, in some parts of the world, an increasing tendency to accept such arguments. You have probably read that the State of California has passed a law that all school textbooks of biology must include an account of Creation as an explanation of the diversity of the living world, alternative to evolution, and, the State legislature claims, equally intellectually respectable. I somehow doubt whether Tony would wish his argument to be pressed quite so far.

However, I think our knowledge of the evolution of language is in a rather similar state to our knowledge of other major evolutionary changes, in the sense that we have very little actual evidence of intermediate stages. I am no expert on comparative linguistics, but from what I am told there are no such things in the world as really primitive languages. Apparently you can say anything equally well in any of the known languages. Some of them may be a bit clumsier in handling certain kinds of subject matter, and others may be especially appropriate in other connections, but all languages seem to be essentially equivalent in their efficiency as means of communication. But to an evolutionist what this implies is not that there never were any intermediate stages, but rather that the steps leading to fully developed languages were passed through quickly, and the intermediate stages have disappeared. I think that Tony finds it difficult to account for the evolution of language because he takes it for granted that he has to account for the evolution of a fully developed language at one step. For instance, he says: 'Whatever it was that the first language-users acquired, it was not a language taught them by their parents'. Because by definition the parents (people *before* the first language-users) could not use language. But surely this is just the old paradox of which came first, the chicken or the egg? I should argue that the parents may have had a primitive sort of language, which the children learnt and improved on; in fact, improved on quite quickly in evolutionary terms, until what had started as only semi-articulate grunts became converted into a fully-fledged language.

I would now like to pass on to some further, more particular aspects of Tony's argument. One point he made was that human language is conventionalised in quite a different way from that in which animal behaviour is ritualised. He argued that when we say that human language is conventional, we mean that many types of behaviour will be adequate, provided they obey certain general rules. We can utter the same sentence using several different modes of expression; not only speech, but writing, the morse code and several others. I should agree that this is true of many languages as they exist today, but I doubt if it is very relevant when we are considering the evolutionary origin of language. After all, writing was certainly invented long after language was first expressed in speech. There are many languages spoken in the world today for which no conventional form of writing exists, and I suppose about half of the world's population is illiterate, although they can all speak some language or other quite adequately. The other modalities, such as the morse code, semaphore and so on, have for the most part been invented only in the last one or two hundred years. I see no reason to suppose that at the time of its evolutionary origin language could be expressed in anything more than the modality of speech (I include in that the bodily gestures that usually accompany speech).

Tony goes on to argue that animal communication is of quite a different kind to language, in that it does not use conventional forms which convert certain things into symbols. He gave as an example the types of behaviour involved in sexual display. However, I think there are other types of animal behaviour which come much closer to using symbolic forms in communication. Consider for instance intra-specific fighting, that is to say fighting between two members of the same species, such as two dogs. It is very common to find that such fights are terminated before any great damage has been done. Possibly the under-dog rolls over on its back and exposes its belly or makes some other similar move, but this is not taken by the upper-dog as an opportunity to get in a really wounding bite, but rather is accepted as a signal that the other chap is throwing in the towel, and calling the fight off. The evolutionary mechanisms for producing such types of behaviour have recently been discussed by Maynard Smith in *On Evolution*. His discussion is in rather elaborate terms taken from Games Theory, but I am not sure whether this is really necessary. I think a main, and perhaps a sufficient, point is that when such a conventional 'throwing in the towel' signal becomes widely accepted, it will be very useful not only to the under-dog, but even to the upper-dog, because after all who can know that the upper-dog today will not find himself the under-dog in another fight tomorrow? I do not see any impossibility in supposing that there could be a gradual evolution of a more and more definite acceptance of various forms of behaviour as symbols in a system of communication, rather than as incidents in a causal sequence of processes. This might apply whether the types of behaviour accepted as symbols were movements such as the dog rolling on its back, or were sounds accepted as words.

It seems to me that human language might have evolved in a way very similar to the evolution of these conventional gestures in animals. We know that early man often hunted game much larger than himself, for instance mammoths. Presumably the men hunted in co-operating groups, and surely methods of communication would have been very useful. It would have been advantageous not merely to make a loud noise and point, but to be able to convey things like 'It's not behind that bush, you silly clot! It's in that bunch of reeds. And it isn't a lion, it's a buffalo.' This would probably require only quite a small vocabulary, and a relatively primitive mental apparatus, but it would definitely be a beginning from which language might have developed, and it would also be related to some of the things which we know that animals can do. Birds of the same species, but from different localities, may sing slightly different songs, which their fellows seem to recognise. Some birds have different alarm cries, which indicate whether the source of alarm is a hawk or a snake. These cries are, of course, not names in a fully symbolic linguistic sense, but they do convey instructions of some moderate degree of precision. As we have seen in previous talks, language is to be thought of rather as instructions than as statements or descriptions, and these types of animal behaviour involve the conveying of instructions and therefore provide, it seems to me, a basis from which language might have evolved.

I admit that the basis for language provided by these observations on animal behaviour is not very sturdy. We can find some clue as to how the symbolic expression of instructions might have evolved, but it is more difficult to see the origin of the structural grammar by which separate instructions are related to one another. We know very little about this, and I rather doubt whether many people have studied those aspects of animal behaviour which look most likely to provide suggestive information. I should like to know, for instance, much more about anything analogous to grammar underlying the structure of communications which a shepherd may have with a sheep-dog. If one watches the dogs being controlled at a sheep-dog trial, it is quite clear that precise and complicated instructions are being communicated, although the communication is non-verbal. It looks, at first sight at least, as though it must have a structure of a kind analogous to grammar. I wonder if anyone has really studied this fully in the light of modern ideas about linguistics. I think it is in this sort of context that one should look for the evolutionary origins of the grammatical structure of language.

Thus, although I agree that we still have an enormous amount to learn about the evolution of language, I do not think it is impossible to get some general clues as to the ways in which it might have happened. Tony seemed to argue not only that we do not know what actually happened in the evolution of language, but that it is in principle impossible to conceive of language as having come into being by a process of evolution. I believe that his difficulty arises because he was thinking of language as having evolved as a single sudden jump, whereas I should argue that it must have been acquired in a series of steps, probably rapid steps, after which man did what Wittgenstein described as 'throwing away the ladder after he has climbed up on it'.

LONGUET-HIGGINS

The question in Tony's interesting talk which struck me as the most challenging one was, where could universal grammar have come from, how could it have originated. I don't suppose he would want to put up a barrier, and say it was no good for us to continue to ask this question, and so I want to try and pursue it for a minute or two.

Quite plainly, as Wad's just pointed out, the grammars of the sorts of language we have now are much more complicated, to say the very least, than the grammars of the earliest languages, and one must suppose therefore that the universal grammar which determines our approach to the noises that our parents make is also probably a good deal more complicated now than it once was. The most simple rule that could have any claim to be taken seriously as a rule of universal grammar, is one according to which any human language must have a phrase structure grammar. We know that human languages have more than just phrase structure grammars, but phrase structure is the most important obvious element in the structure of all human languages. In the book by Miller, Galanter & Pribram called *Plans and the Structure of Behaviour,* the authors consider how we might try and assign structure, in an abstract sense, to different sorts of behaviour. They make a very plausible case for saying that much of our rational behaviour has a structure which is like phrase structure. Let me try and indicate what I mean. In making a plan, one may decide to do A and then B in order to carry out the plan. But in order to do A, one has to do X and then Y, and to do B one has to do

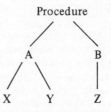

Z; so if one is going to carry out the whole plan one first does X and then Y, and then one turns to the next stage in the plan and proceeds in the same manner. Now, a linguist will recognise that the order X, Y, Z which arises from this structure (see figure) is closely analogous to the order of the words in a structured sentence. Now, I'm not suggesting for a moment that our languages as we find them are simply characterisable by saying that they have phrase structure, of course not; but none the less, the most up-to-date theories of language take phrase structure as the basic component of the grammar, and any subsequent transformations are applied to this structure. I just wonder whether along such lines we mightn't hope to understand a little more about the origin of universal grammar. Perhaps universal grammar took over for symbolic communication a set of ways of thinking which were already useful in the formulation of plans; and presumably the ability to formulate plans must have considerably predated the appearance of language as we know it.

LUCAS

I'm going to find myself in the position of being much more of a naturalist than Tony. His position depends on a false dilemma between the language which he says is constituted by conventions, and the other source of signs which are based not on convention but on displacement activity; and he wants to drive a very sharp wedge between these in order to produce a gap over which evolution cannot leap. But conventions can be natural; and if you start by having a convention which is natural then it can become more and more conventional thereafter. For instance, we often now in university circles see languages in very abstract terms. But if you come next year to read what we've said today and read it in cold print, all the words will be there but you'll find it's very heavy reading. You depend immensely for understanding on context; in our case the way our hands work and all sorts of other things, in order to get the full sense of what we are saying. Even in our sophisticated languages we are still very much dependent on a large number of other forms of behaviour which are only partly conventional, and partly natural. Or take a second example: earlier today I was in Edinburgh Castle which was redolent with the symbols of Royalty — Crowns, Thistles, Orbs and Sceptres; there was a Sword of State, and this now is a *convention* of the civil power; but once it wan't a convention — it was a real threat; when Bruce wielded that sword, you obeyed. Even now the Keepers of the Castle recognise this point, and next to the Crown Jewels of Scotland, they have the RAF room, and the battle-colours and the guns.

This explains how the jackpot phenomenon could have occurred. You start with various *natural* conventions, the dog giving his underflank to indicate its willingness to be killed — this is absolutely natural — and then because it is natural, it becomes understood, and because it becomes understood, it becomes in its own right a symbol. It seems to me perfectly unexceptionable that some account of the sort that Wad has given should bridge the gap between what are at first obvious responses to the situation, a purely natural form of activity, and later, more conventional, behaviour. We move in our own society in a very similar way — from the original generosity to the conventional glass of sherry or conventional hand-shake; when I shake your hand, it no longer means that I haven't got a sword in my hand, and so, that I will not kill you — it is an entirely conventional sign.

This is one of the difficulties that Tony put forward. There's another one, however, which was about how to get an abstract grammar, and this is largely answered by Wad who points out that you can start with a much lower-grade one, and then work up to the abstract one by having a series of translation rules — you can go from speaking to writing, to Morse, to all sorts of other symbols, and this is what underlies the concept of abstractness. Then there's one other biological point, namely that even the birds, I'm credibly informed, have some sort of abstract grammar. Konrad Lorenz and the other ethologists brought up some birds in human company and the birds made a great mistake; they thought that they were human beings, and that Professor and Mrs Lorenz were their parents. It is clear that in this case the birds had an innate abstract grammar in the sense of a correlation between parents and the appropriate sort of behaviour, which has then been filled in with the wrong filler. The birds' mistake lay in their specification of this abstract grammar into a particular one, due to their misidentifying their parents. They therefore tried to feed their parents with worms — and as they found they couldn't get them into their mouths, they put them into their ears.

KENNY

I don't want to claim, and I don't think I did claim, that it is in principle impossible to explain the origin of language by natural selection. I said that there was a difficulty of principle which had not been faced up to, that is, a different difficulty from the general difficulties which Wad reminded us of, that crop up all over the evolutionary scale. In the case of the origin of the birds, what we are lacking is knowledge of the existence of the intermediate cases which the theory would postulate. The special difficulty with regard to the origin of language, is the difficulty not of providing an intermediate case, but conceiving exactly what *is* an intermediate case between linguistic and non-linguistic behaviour. In spite of the erudition and ingenuity of the other symposiasts I'm not convinced that this evening produced something which is at all like an intermediate.

I don't deny that animal systems of communication are

very efficient in communicating a lot of information to one animal about another, but you can do that without having a language as you can best see if you study human non-verbal communication of the kind which John mentioned. If you overhear a conversation in a neighbouring railway compartment in a language of which you don't understand a word, you can get a lot of information about the age, the sex, the temper, and so on, of the people. You may realise, for example, that there is an old woman who is very angry indeed with a small child who is being very naughty. This is the sort of thing which is a close parallel to the information that is communicated in the bird's system of communication. The examples, the human analogues which John gave were, it seems to me, all non--linguistic cases. If one is going to use the word 'grammar' so broadly that even putting worms in somebody's ear is an exercise of a linguistic ability, then all serious discussion of language must come to an end.

Eighth Lecture. Possible Minds

Now that we have done our best to expose the severe limitations on our present understanding of the phenomenon of mind, perhaps I may be excused for throwing academic caution to the winds and arguing from what is to what might be. I want to begin by asking whether, and how, modern science has fundamentally changed our thinking about ourselves and our relation to the cosmos. Allowing that it has, and that for better or worse these changes cannot be reversed, what does our new world view imply about the future development of humanity and the possible existence of intelligent life elsewhere in the cosmos? And finally, whether or not such speculations seem to be well founded, how does our present image of man as an intelligent being affect the traditional notion of purpose, as applied both to individual human beings and to the universe as a whole? To let one's hair down in this way is to incur the risks to which Teilhard de Chardin exposed himself in his book *The Phenomenon of Man*; all I can hope is that some of these issues would have been deemed by Lord Gifford to be relevant to Natural Theology.

Christopher Hill has recently issued a warning against buying total history from partial historians, and the history of science can be a dangerously partial study. But even the Marxist historian, who insists that the development of science is driven by economic pressures, cannot ignore the extent to which each of the great scientific revolutions has altered man's view of himself. The Copernican revolution placed him in a much bigger universe than before, though not at its centre. The Newtonian revolution established the universality of natural law, so that he could feel secure against the whims of arbitrary forces. The Darwinian revolution placed him at the peak of a vast evolutionary development, and post-Darwinian biology has shown him where to look for the secrets of his inheritance. Relativity and quantum mechanics have established him as the observer who cannot be omitted from any complete account of matter and motion; and the computer revolution has enabled him to begin to construct working models of his own cognitive processes. Whether, as the Victorians confidently predicted, this increase in scientific understanding has been accompanied by steady progress in human well-being, is open to question. But it could hardly be denied that twentieth-century man has a model of the cosmic machinery which is incomparably grander and more comprehensive than that of his great-grandparents.

At this point the question inevitably arises; does any amount of machinery make a machine? Or is the cosmos rather to be compared with one of those assemblages of bolts, levers, crankshafts and pressure gauges that adorn so many exhibitions of contemporary sculptures? Let us pursue the comparison for a moment. It could be argued that what distinguishes modern science most sharply from its predecessors is its internal coherence. We are no longer prepared to regard the occurrence of an earthquake, or the appearance of a comet, as an intervention by some cosmic artist bent upon relating the nuts and bolts of human history to the pressure gauge of the earth's crust by artistic principles beyond our comprehension. We expect a scientific account of earthquakes to relate their position and intensity to such things as the stresses in the interior of the earth and the presence of faults in the overlying rocks. And by and large such accounts are forthcoming. The geographer turns to the geologist, and the geologist looks to the geophysicist, for information that will help him make sense of the seismic data and, with luck, to predict the location of future upheavals. The aesthetic principles which unite the various bits of cosmic machinery are not arbitrary and whimsical but surprisingly coherent and intelligible; far more intelligible than we have any right to expect. The pursuit of science — and a very rewarding pursuit it is — abundantly justifies our implicit faith in the ultimate intelligibility of nature. For whatever reason, the universe in which we live seems to have a great deal more internal logic than a mere assembly of spare parts — if such an understatement were not an insult to such a beautiful creation.

In using the word 'beautiful' I think I speak for the vast majority of people who have worked at the frontiers of science. No-one who looks closely at nature can fail to be moved by her austere beauty, or to feel a sense of inevitability that this beauty would one day be beheld by her children. One feels privileged to possess the equations which describe the motion of every drop of water in the ocean, and to know that the forces which hold the moon in the sky are exerted by the remotest masses in the universe. We now know too much ever to believe again that the world is a mere jumble of events of which we are the helpless victims.

Having, I hope, generated some feeling of surprise that we should be able to understand the world even as well as we do, and that its investigation should be so deeply satisfying intellectually, I will now move on to consider whether this is merely a contingent fact about ourselves, or whether it is a manifestation of some phenomenon which we might expect to occur elsewhere in space and time. From our own point of view, certainly, twentieth-century science is a very special phenomenon, unprecedented in the history of our species and unparalleled in the development of any other, as far as we know. Modern science seems to be a clear case of 'hitting the jackpot', to borrow Waddington's happy phrase. It could hardly have been foreseen that such a straightforward idea as that of the controlled experiment would lead so far in such a short time and give such power over our material environment. But there is something about our science which seems to differentiate it from biological inventions such as the elephant's trunk. It is not just useful — it seems to be *true*. I am not of course denying that much of science is provisional, or claiming that all its present concepts are sacrosanct; in many ways it is a very imperfect edifice. But I could not concede without cowardice or dishonesty that there might be as much truth in astrology as in modern astronomy, or in necromancy as in the social sciences. In science as in scholarship generally, one aims to get the *right* answers to one's questions. There will admittedly be differences in style between different formulations of the same knowledge; but nothing can alter the content of Mendeleeff's discovery that there is a periodicity in the chemical properties of the elements, or the fact, established by Watson and Crick, that the two strands of DNA are complementary, with all that that implies about biological replication.

One could argue, as Eddington did, that the laws of science are ultimately of our own making, and that is why we have the impression that the universe is governed by immutable and universal laws. Eddington's argument has considerable attractions, as being the only attempt by a modern scientist to account for the existence as well as the content of scientific laws. The laws of physics, he said, arise ultimately from the way in which physicists measure things; and he proceeded to derive the properties of protons and electrons from certain basic principles of measurement. Unfortunately he failed to predict anything that was not known at the time, and much has been discovered since, so his ideas are no longer taken seriously by professional physicists. But in retrospect it is difficult to see how Eddington's principle could ever have sufficed to account for the extraordinary reliability of natural law, not only in scientific laboratories, but also when no physicist is around to check that every proton and every electron is conforming to his methods of measurement. One cannot escape the feeling that there really *is* a natural order, and that at last, after thousands of millions of years, one of the creatures of that order has awakened to its existence and has penetrated with his imagination not only the remote corners of space and time but even the atoms and molecules of which he himself is composed. Let us turn a blind eye, for the moment, to his terrifying moral immaturity, and ask whether his existence is of any significance at all.

In an earlier lecture, when we were discussing the evolutionary development of mind, I remarked that once evolution had got under way it was pretty well bound to produce something

110

interesting, and we like to think that we are specially interesting. The remark was, of course, deliberately paradoxical in that for something to be interesting there must be someone to take an interest in it, and the existence of such a person is at least as interesting as those things in which he is interested. So we really ought to be asking the thoroughly scientific question whether the existence of intelligent life is unique to this planet, or whether its appearance is a general phenomenon, which we may therefore expect to have occurred elsewhere in the universe. In our present state of ignorance we are in no position to answer this question, though serious scientists have actually begun to scour the heavens for evidence on the matter. Can we stretch our imaginations enough to think of the forms that such intelligences might take, and to hazard predictions about our own descendants on this planet?

The creation of imaginary worlds inhabited by imaginary beings is a literary tradition which has been honoured by writers as respectable as Homer and George Orwell; it is now fully accepted as a vehicle of social comment. So I will single out some of the recurring themes of this tradition, where they seem to touch on the question of possible minds. Let us start with one which often sends a chill down the spine, namely the thought of 'alien life forms' as we might meet them in the pages of H. G. Wells or on the cinemascope screen. The Bug-Eyed Monsters and the Little Green Men horrify us on first acquaintance not so much because of their twisted minds but because of their hideous bodies. Only when we see such monstrosities do we realise just how much we take our own bodies for granted as the miraculous pieces of engineering that they undoubtedly are. It is a besetting sin of the academic to dissociate the life of the mind from that of the body which sustains it. One can all too easily forget, when contemplating the nature of intelligence, how profoundly our modes of thought and perception are dependent upon the sense and limbs with which nature has endowed us. When, in *The Merchant of Venice*, Shylock is moved to protest his humanity, his first words are not about his mind but about his body:

'Hath not a Jew eyes? hath not a Jew hands, organs, dimensions, senses, affections, passions?'

It could well be supposed that the technological spree on which humanity has recently embarked is likely to come to grief unless we can learn to respect our bodies as much as the minds of which we are so inordinately proud. The physical performances of running over rough ground, or even of untying a parcel, are at least as remarkable as the ability to solve quadratic equations, at which computers beat us hollow. In many ways our most remarkable faculties are not those for which we claim special distinction, but those which we share with the humble beasts.

Another idea which haunts the imaginative writer is that of

111

the artificial man, be it a Frankenstein, a Universal Robot or Stanley Kubrick's computer Hal, which is capable of taking command of an expedition to Jupiter. In the pages of science fiction the artificial man appears either as a helpful assistant or as a threat to the human race, according to the design of the computer program that controls it. What might count as virtue or ambition in such a being is assumed to be clear from our own experience of such qualities. The artificial man is none other than our own image reflected in the distorting mirror of modern technology. Its bulging brain-case accommodates an intelligence which computer scientists burn to emulate; an intelligence pure and logical, to which any concern with human beings and their tiresome affairs is at best a regrettable necessity. The means whereby the artificial man is to perceive the world, to interpret the behaviour of non-artificial people, and to behave sensibly and gracefully, are of course matters on which much expensive research urgently needs to be done. Until the results are forthcoming we shall not know how soon the human race is likely to become dispensable, or what sort of mind will be ruling this planet after we are gone.

A modified version of this fantasy is that in which the minds of the future are only semi-artificial. In a recent enquiry addressed to members of a scientific club I belong to, an influential American research agency asked for views about the desirability of constructing information-processing systems which would incorporate both computing equipment and human beings, whose brains, it was recognised, were in some ways superior to existing computers. To allay any possible anxiety the assurance was given that no plans had been contemplated for making direct electrical connections between computers and human brains. Here again we have an area which is clearly ripe for technical exploration, though my own feeling in the matter is that the most reliable input to the human brain is through the human senses, and the most informative output from it is what people say and do. We are free to form our own opinions on the ethics of experiments involving 'human' nervous systems; all one can hope is that people will not be so foolish as to do experiments which are not only inhumane but intellectually misguided into the bargain.

An altogether more engaging line of speculation about the development of mind begins with the thought that human beings are at their best when communicating with their fellows − that indeed no human being is complete on his own. Human society is compared to a multicellular organism, in which the individual cells are supported and restrained by one another. May we not look forward to a further step in evolution, to an enhanced intimacy of communication which creates, in some real sense, a collective consciousness? Already, with television and mass transport, we can look in on people all over the world and begin, or so we think, to see life through their eyes as well as through our own. Need there be any limit to this increased awareness of other human beings, and of the astonishing universe in which we all live?

The concept of the corporate society is as old as philosophy, and has been the cornerstone of some of the great religions. No-one would dare challenge it as an ideal, though some people are prepared to forgo its standards in the means which they recommend for its ultimate achievement. The corporate society raises, however, the central question of political philosophy, namely the right relation between the welfare of a society and the wants of the individuals composing it. In the human body each cell must be subject to the laws of the whole, or the organism will sicken and die. But is the individual human being to be sacrificed to the demands of the state? Common humanity seems to dictate otherwise. As yet human societies are not genuine organisms; the things we value most in our civilisation are not the pyramids but the achievements of free men, and we reserve the right, as individuals, to pass judgement on the societies of which we are members.

So we have to ask: are the new powers offered us by technology really helping us to move towards the ideal society in which we can speak without metaphor of a common mind? It is, of course, early to say; but of certain dangers we can already be aware. One danger is that of crudity in our judgements on other people, their actions and their achievements. With so much information pouring in about the mistakes, misdeeds and misfortunes of complete strangers, we are tempted to take the latest news at its face value, and to accept uncritically the snap judgements which go with hastily delivered reports. Would we want other people to sum up our problems and propensities as hurriedly as we are prepared to sum up theirs? I doubt it.

Another hazard which besets the mass media is what I will call one-way communication. If I witness a road accident on the A1, I may be in a position to help in some way – perhaps by administering first aid or telephoning for an ambulance if someone is seriously hurt. But if I see on the television some horror scene from Biafra, Belfast or Bangladesh, there is little or nothing I can do to mitigate the pain of the victims. Repeated experiences of this kind can hardly fail to dull my sensitivity to the misfortunes of others, or worse still, to replace a genuine compassion by a morbid and irresponsible curiosity. One cannot maintain for long a sense of responsibility if one lacks the power to fulfil it.

But even in morally neutral circumstances one-way communication may dull our sensibilities while seeming to enhance them. To a musician one of the most rewarding of musical experiences is to play in a group, with each member adapting his playing to that of the others, so that the performance approaches as nearly as humanly possible the product of a

113

collective mind. A few years ago I had the unsettling experience of playing with a violinist whom I could simply not keep up with; he played faster and faster until we finally broke down and he then had the audacity to ask me why I was hurrying so. It turned out that he had been practising with a record of the piano accompaniment, and I suddenly realised that he must have formed the habit of playing slightly ahead of the record. This did not matter so long as the accompanist was deaf to his playing, but proved quite disastrous in a situation demanding genuine musical co-operation. Or, to take another example, rather nearer home: even such an innocuous medium as the formal lecture can fail to spread enlightenment if the lecturer is unable to read the faces of his audience. One need not go so far as to suggest that one-way communication is invariably worse than no communication at all; but one may well doubt whether it will bring us much closer to that community of mind which we would like to think of as a characteristic of the ideal society.

Let me try to draw some of these threads together. Whatever one's guesses about the other inhabitants of this universe, or about the quality of their minds, we cannot doubt that the invention of science has been a great intellectual leap by our own species, perhaps an even greater one than the invention of language, upon which the pursuit of science obviously depends. In so far as science suggests to us the forms that other intelligences might take, or the kind of society that our descendants might create, it holds up a mirror to our own nature and warns us not to ignore our physical character, or put too much trust in technology, in planning the brave new world of tomorrow. We cannot yet tell whether the emergence of mind is a general phenomenon of which we are merely one instance, but if it is, would we be able to give a good account of ourselves if summoned to judgement by wiser beings than we are? And this raises the last question I would like to discuss: does our new knowledge of the world, and our new power over it, enable us to see any more clearly where we ought to be going, or how to get there?

I am not one of those who believe that we can learn moral lessons from evolution, or that human beings are under an obligation to bow to man-made laws of history. But there is a principle which we cannot afford to ignore, and which has actually been elevated into a moral principle in the Islamic religion, namely the Law of Cause and Effect. Whatever one's ethical system it will be of little use unless one can predict the consequences of different courses of action. One thing science has done for us has been to increase by an order of magnitude both the range of possible human objectives and our knowledge of how to achieve them. And at last we are beginning to understand ourselves properly, and to see the need to match our morality and our technology to the minds and bodies with which nature has equipped us.

But these remarks do not directly address the question whether there could be said to be a cosmic purpose, to which human goals and plans should conform. To such a difficult question it would be absurd for me to offer an answer. But perhaps the question can be replaced by a more practical one: if there were a cosmic purpose, how could it best be discovered? And here perhaps is where science really comes into its own. Because if this most recent product of the human mind has any value beyond the power which it gives us to push things and people around, it must surely be seen as an opening of our eyes to the true nature of the universe and our relation to it. In order to discern a cosmic purpose we must be able to see what the world is really like, though at any particular time our vision will inevitably be clouded to some degree by ignorance and prejudice. Perhaps when the vision is clearer and brighter our descendants will grow into such an intimate relation with the universe that their own purposes become merged with those of the world whose consciousness they embody. We ourselves must be content with the thought that it is better to travel hopefully than to arrive.

Discussion
KENNY
I think we must all be grateful to Christopher for the elegant way in which, in the first part of his paper, he displayed the connections between scientific truth and cosmic beauty. Truth is not beauty and beauty is not truth, but the two concepts are more closely connected than is allowed for in many philosophies of science. This is shown, among other things, by the use made in choosing between hypotheses of the aesthetic criteria of elegance and simplicity.

I would maintain, without attempting here to prove, that any intelligence, human or otherwise, which was capable of possessing the truth about the universe of science, must also be an intelligence capable of appreciating its beauty. But in a world in which there are no intelligent beings, there is neither truth nor beauty. If there is an intelligent and eternal God, then of course truth and beauty are equally eternal. He knows, and always has known, all truth and delights and always has delighted in all beauty. But if there is no God, then I do not see how to make sense of some of the remarks which Christopher made in his lecture. I do not see for instance, how there could be aesthetic principles connecting together the various bits of cosmic machinery. I do not see how there could be any cosmic purposes to which our own purposes may eventually become merged. It may be, I do not know, that when Christopher was expressing agnosticism about the existence of cosmic purposes he was equally expressing agnosticism about the existence of God. It does seem to me that cosmic purposes without God are nonsense.

I'd like next to defend the unique interest and importance of intelligence against Christopher's rather hearty remarks about the importance of the body and the need to defend it against academics. The physical performances, he said, of running over rough ground or even of untying a parcel, are at least as remarkable as the ability to solve quadratic equations at which computers beat us hollow. I must repeat that computers beat us hollow at solving quadratic equations only in the sense in which clocks beat us hollow at telling the time, but let's waive that. The physical performance of untying a parcel is not a remarkable performance by the untier. If it shows anything remarkable it shows something remarkable about the creator of the untier, whether that creator was God or nature, or both. The reason why Christopher is impressed by it is that he thinks to himself: 'What a mind somebody would have to have in order to be able to construct a robot to untie a parcel'. What really impresses him is the mentality latent in the physical procedure involved. It is really the mind and not the body of the parcel untier that is impressing him.

I turn to a more serious point — Christopher asked whether the new powers offered us by technology are really helping us to move towards the ideal society. He had misgivings which I share, but I have others which I'm not sure whether he shares. He spoke of the dangers of one-way communication. It may be true that there is little that one can do to mitigate the pain of a particular sufferer in Belfast or Bangladesh, but of course it's quite wrong to suggest that there isn't anything one can do to mitigate suffering in Belfast and Bangladesh in general. At least there is always political action, there is always money that one can give. These activities, of course, don't have the immediate appeal or satisfaction that the giving of first aid has on the A1, but it's surely wrong to pretend that technology here doesn't increase our ability to do good. But *eo ipso* of course, it increases our responsibility for evil, and that is the main point which I want to develop.

The Pharisee who passed by the man who fell among thieves was blamed because he knew about the victim's sufferings and he could have helped him as the Samaritan did. If the Pharisee had been in the synagogue at Capernaum, a long journey away by donkey from the Jerusalem-Jericho road, no-one would have blamed him. Nowadays, the Pharisee can see the victim on the television and he can get from Capernaum to Jerusalem by jet. That the misfortune which the Pharisee sees on television is a stranger's misfortune seems to be neither here nor there.

I would wish to claim that it is not an accidental fact that there is a tendency of science to corrupt — and I mean this absolutely seriously. The way in which I'd like to prove this is very simple. Science gives power, and power corrupts, and therefore science corrupts. It is first of all obvious that science gives power. As Christopher himself said, one thing which science has done is to increase by an order of magnitude our powers of achieving human goals. That power corrupts is usually understood as meaning that if you have power this enables you to do evil: royal power may give you the power to cut off people's head and send them to prison. That was the kind of power that Lord Acton had in mind when he framed his aphorism; he meant that if you have power to do evil, human nature being what it is, the likelihood is that you will sooner or later succumb to the temptation power brings and do evil. It seems to me that at least as important a corrupting feature of power is that it gives you the ability to do good and therefore that it puts sins of omission as immediately and inevitably within your grasp as it puts sins of commission.

If one accepts two fairly plausible principles, it seems easy to show how power corrupts. The two principles are these: first, that the more evil you are responsible for, the more evil you are, and second, that we are responsible for evils which we know about and which we can prevent without disproportionate loss. Now as science increases our knowledge of the world, it increases our power; it increases our power for removal of evils, and to the same extent it increases our responsibility for them.

Now all the stages in this argument are too simple, but if they were restated with due qualifications, so that they were acceptable, I think they would still lead to the conclusion that other things being equal science corrupts, that is, the technological power that science brings makes one responsible for ever greater evil. Other things being equal that is what science does.

What are the other things that might not be equal? The most important is human motivation and, in particular, human selfishness. We might take as a simple and crude measure of selfishness — in contrast to altruism — the amount of one's time, energy, money, power, that one devotes to the satisfaction of one's own needs and desires, and the amount of time that one devotes to the satisfaction of the needs and desires of others. Let us imagine an average selfish man, *l'homme moyen egoiste*. Let us suppose that he devotes half his resources, half his time, money, energy and power, to himself, and half his resources to others. And let us consider the process to which Marx drew attention, by which as technology advances we devote less and less of our working life to actually providing ourselves with food, warmth and clothes.

In a primitive society the average selfish man is free from blame. It takes, let us say, half his working life to support himself; he devotes half his working life to supporting himself, and he devotes to his fellow men all the time that he has to spare from keeping himself alive. He's not responsible for any

117

of the human suffering around him because there is none that he can remedy without imperilling his own life — something which is maybe estimable and heroic to do but which is not normally obligatory to do.

But in a highly technologically developed society like our own, the position is very different. Many people have an income more than twenty times as much as is necessary to keep them at subsistence level. Such a person, therefore, has earned enough to keep himself alive for the week by the time he reaches the coffee break on a Monday morning. Now, what is the moral situation of the average selfish man in this situation? As long as there are, within his knowledge and ability to help, other human beings below the subsistence level, he shows up very badly indeed. He needs only to devote only 5% of his resources to keeping himself alive, and in fact, he devotes 50% of his resources to the satisfaction of his own desires. Over 45% of his activity expresses a preference for superfluities for himself over subsistence for others. On the account of responsibility which I gave earlier, he is responsible for the amount of suffering that could have been prevented by his allocation of that 45% of his effort to altruistic purposes.

Once again, the model that I have presented is greatly oversimplified — it's extremely crude; but once again, I believe that if the simplifications were removed, the conclusion would still be warranted that technological progress is bound to corrupt unless altruism increases *pari passu* with scientific knowledge and the technological power it gives. This is true at least in those circumstances in which scientific progress has so far been made, namely, where a significant part of the human race lacks the necessities of life. Sadly, there seems little reason to believe that there is a positive correlation between altruism and technological power. I find it hard, therefore, to share the optimism of Christopher's conclusion — I find it hard to look forward to the cosmic harmony to be diffused by the progress of science. Similarly, I am not as depressed as Christopher would be by the prospect that John Lucas holds out that there may be areas of human behaviour to which scientific explanation can never penetrate.

Christopher pointed out quite rightly that the beginning of science was as important an intellectual leap in human history as the origin of language. Perhaps it was the origin of science and not the origin of language that was the tree of knowledge which brought our fathers' fall.

LONGUET-HIGGINS

I do agree with very much of what you've said. I was trying to make a sharp distinction, which of course is difficult to maintain in this greedy world, between science, which is the disinterested pursuit of the understanding of nature, and technology, which is motivated by more material considerations. I was trying to express my distrust of technology,

118

although just to throw it away is to shirk the additional responsibility which it brings and I couldn't agree more that this responsibility is inescapable.

But can I just return to Tony's very first point, which was about my position committing me to theism. If I knew what sort of deity he might be discussing, I think I would be prepared to say whether I believed in such a deity or not. I was trying to convey the feeling that comes upon anyone seriously engaged in attempting to understand nature. The feeling that when one discovers something, it was sitting there waiting to be discovered. One gets into the habit of expecting to feel that way, and by and large one's expectations are fulfilled. If one didn't have the hope that next time one obtained an insight, it would give one the same feeling of inevitability as the last one, one would probably give up research, or stop having ideas. Now, whether that constitutes a belief in God is for you, I suppose, to decide rather than for me, according to how you define your terms.

LUCAS

First of all, I want to defend Christopher's optimism against the apparently rigorous proof given by Tony of an almost inevitable selfishness. It seems to me to rest on a false dichotomy — we would not be better off if the skilled surgeons in the many hospitals of Edinburgh by the coffee break on Monday morning decided they had earned enough for themselves and now should start working for others. That is to say, there is not — and this is one of the fundamental facts of our social life — a sharp dichotomy between working for oneself and working for others; the same activity can be *both* for you *and* for me; on this fact Christopher can reconstruct the political theory that he didn't actually give us, so as to be immune from Tony's criticisms.

What I would now like to do is to pick up the lead Christopher offered at the very end where he invited Tony to provide him, as it were, with an emporium of possible gods, and then, as he was introduced to each of these potential deities, he would pronounce: 'I'll have this one but not that'. This is a mistaken approach. The correct way is for Christopher first to think about the nature of the mind and the nature of the universe, and try and find words which express this; and *then* face Tony's needling — 'If you say of truth, and of beauty too, that they are not merely temporally conditioned but fundamental features of the universe, are you not committed to *something*? — not perhaps the God, that you don't like, of the Westminster Confession — not of course, the God of the Thirty-Nine articles, not even the God of Vatican Two, but *something* other than oneself making for good, making for beauty, making for truth?' I think this is something which Christopher is committed to. As he developed the various natures of possible minds, certain marks of what it was to be a mind emerged. It was essential that a mind could reserve the right, as an individual, to pass judgement. Minds were intimately connected with not merely utility — the sort of things that computers can do — but truth; science was important not just for being useful but because it seems to be true. These are two of the marks of mind; and a third point emerged, that it was the nature of minds to be able to share. There is something wrong with one-way communication. Communication is properly a two-way process, and thus one has individual minds starting to form a common mind. These are three marks of what it is to be a mind. Christopher then began to find that he wanted to impute them to the cosmos as a whole, not perhaps in our generation, but at least in those of our children or our children's children; they would be able to merge their purposes with the cosmic purpose. And having said that, Christopher has thereby disposed the onus of argument that way round — he can't go and say: 'I don't want this or I don't want that, you name it, Tony, and I'll disbelieve it'. Rather, having committed himself by describing the nature of mind and the nature of the universe as best he can, he has to expose himself then to Tony's probing, to show what sort of things are presupposed if these views are themselves to be rational.

WADDINGTON

It seemed to me that one of the main points of Christopher's talk was his conclusion that the universe does have a structure. We usually express this as a causal structure, which we expound as the laws of science. These are obviously to some extent moulded by our sense organs and our particular type of intelligence, but as Christopher pointed out, we do not invent the laws of science out of the whole cloth. To a large extent they reflect the structure of the world which we have to deal with. I think it is a very basic point that science is the main method we use to discover something about the structure of the universe.

I should agree with Tony that knowledge of this structure brings power. But I am afraid I was completely unconvinced by the argument he based on the idea that power corrupts. That phrase was a rather snide remark by a Victorian politician, who was really talking about political power in the corridors of Whitehall. I don't think it can be generalised to apply to all power under all circumstances. Tony did just let drop the remark that power also makes it possible to do more good, but then by some very peculiar arguments he tried to convince us that the power to do good in some way makes things worse. This was the point that I confess I find very difficult to grasp.

I should like to go back to Christopher's point that the world has some sort of structure. Any mind on this planet, or on any other planet or wherever you might discover it, would only qualify as a mind if it does succeed to some extent in reflecting this structure. Other minds might, of course, appreciate the world in a very different way from ours. Our minds and the structures we discern in the world are very largely based on the use of the sense of sight. A dog, who relies much more on the sense of smell, must have a very different way of approaching the basic structure of the universe; but if it is going to operate at all successfully, so that its mental activities can qualify as some sort of a mind, it will have to discern the same basic underlying structure, which seems to be there in the universe, and which we can discover to a greater or lesser extent, but which we cannot essentially alter.

This leads on to the questions about God and cosmic purpose. If one thinks of the structure of the universe in terms of simple Newtonian things pushing and pulling each other about, then it would be difficult to suppose there was anything like a purpose inherent in it. In so far as such a world could manifest a purpose at all, this purpose would have to be built into it by some purposeful creator who had made it. But I think we have all been arguing that the world and our perceptions of the world are not really like that at all. We perceive the world through an apparatus which involves our own mental abilities. Our perceptive apparatus could not function unless it had some internal properties of stability, which make it possible for us to recognise something when it comes into our experience for the second time. As I argued at an earlier lecture, these self-stabilising properties involved in perception are very similar in kind to purposes. Now when we speak of a structure of the universe, we mean of course a structure in the universe as we perceive it. That is to say, the components of this universe are not simple material bodies, quite independent of ourselves, but are the types of things we perceive with this apparatus which involves properties similar to purposes. In these circumstances it seems to me possible to say — if you like to use this type of terminology — that the structure of the universe involves a cosmic purpose. I think it is very much a matter of terminology whether you use expressions like that or not.

I think it is also a matter of terminology — a matter of taste, if you like — whether you want to go further and say that if there is a cosmic purpose this implies some sort of a god. But I should wish always to keep in mind the basis from which it seems to me legitimate to derive language of this kind. I mean that the use of phrases such as cosmic purpose or God seems to be derivable from the basic consideration that our knowledge of the world depends on a process of perception which involves something similar to purposes. This is not by any means the only sense in which these phrases are used. For instance, God may often be used to mean a creator who injects a purpose into an existing but meaningless material world. Words like God and cosmic purpose have in fact been used in so many different senses, that I personally tend to avoid them. Some of their uses I am not at all attracted to, though others have much more meaning for me.

KENNY

I don't believe that the power to do good makes things worse — makes the world a worse place. I think that the *unused* power to do good makes *us* worse, makes us worse men.

LONGUET-HIGGINS

Well, Wad has almost said what I would have said myself, I think, because perhaps the distinction of most interest between the conventional theistic position and any other position, atheistic, agnostic, or what you will, is really in whether one thinks of the universe as having been made by somebody outside it, or whether you think of it as containing within itself all the matters of significance to us; and I personally take the latter view. I think that when I spoke of our descendants' purposes being merged with those of the cosmos whose consciousness they embody, I was really thinking of life growing outwards, as it were, until the universe was in such a tight intimate relationship with itself that you could think of the whole thing as a living organism. This is, of course, a wild mystical fantasy borrowed from Olaf Stapledon's *Last and First Men,* but never mind that. Anyway, I would regard it as a confession of intellectual defeat to say that we must think of the universe as having been made by somebody outside it, who had his own purposes which to a certain extent are no business of ours. But, of course, one finds oneself here in such deep water that perhaps the only thing to do is to shut one's mouth.

Ninth Lecture. The Genesis of Mind

The other three lecturers, in the course of their final lectures, have each indicated a certain urge to rewrite the book of Genesis — for different reasons. Waddington wanted to say that the important thing was what he said to her and what she said to him, because this brought about language and that was the cause of original sin. Kenny was also interested in language as being the one thing which evolution could not explain and the one point where, since language was what made man in the divine image, some further explanation needed to be sought. And Longuet-Higgins wanted to change the botany of the Garden of Eden, with the apple-tree being the apple-tree of scientific knowledge rather than the knowledge of good and evil.

Tonight I want to go into these themes rather more fully in both a traditional and a contemporary frame of mind. Traditionally, I want to take up a debate which took place in 1860 between Bishop Wilberforce and Thomas Huxley, which was posed by Wilberforce in the terms of the descent of man: 'Are we descended from apes or angels?' It is generally reckoned, though as a matter of historical fact incorrectly, that Huxley beat Wilberforce, or, in the terms suggested, that 'the apes had it'. Nevertheless, some obstinate questions still remain. We are quite clear now that man is a primate and has evolved from primates who themselves evolved with the vertebrates and the chordates and so on, and if we are concerned only with the genetic explanation, then that is the explanation we must give. But still we feel that something is left out, and this is something I want to bring out today, partly to pick up two other themes which we have been developing in the course of these lectures: the theme of there being different types of explanation, and the theme of autonomy or freedom.

In order to put the question in more contemporary terms than those of a debate 100 years ago, I shall pick out the difficulties as they have presented themselves to Jacques Monod. In his book *Chance and Necessity,* he expressed in modern form many of the difficulties that we feel when we turn to consider the Genesis of Mind. Before I criticise him, let me first praise him. Praise him for his style, praise him also and more significantly for the biblical fervour with which he writes. We keep on hearing about the Old Covenant being annulled and the need for a new one, and this is symptomatic. Monod thinks that he writes as a molecular biologist, ruminating on the implications of the genetic code; but he feels as a man, coming to terms with the human predicament. The reason why his book has caught on is not because of the many true things which he, as a scientist, tells us which we didn't know before, but because he is expressing things which we often have felt but find it difficult to put into words.

The key error which I shall elaborate in Monod's book lies in the title – *Chance and Necessity*. He thinks that these two concepts are opposed, and that you can discover what chance is just simply by contrasting it with necessity. He does not realise that necessity is itself a highly ambiguous word which has many different meanings. Just take one quotation:

'There is no chemically necessary relationship between the fact that β-galactosidase hydrolyses β-galactosides, and the fact that its biosynthesis is induced by the same compounds. Physiologically useful or "rational", this relationship is chemically arbitrary – "gratitous" one may say' (p. 78, cf. pp. 135–6).

Throughout this book, as one reads it, one needs to underline these words 'arbitrary', 'gratuitous', 'chance', 'necessity', 'random', and ask exactly what they mean.

You remember that after the first lecture, when I was arguing against Longuet-Higgins, I produced the case of the pin-table and the result, which looks rather like a cock-hat, of what happens if you drop balls from the top through a large number of pins into a number of slots (p. 10). Monod gives another metaphor, which I shall elaborate as being more congenial to Longuet-Higgins's heart. If I took this microphone and put it too close to the loudspeaker, I should get a positive feed-back; any noise coming into the microphone is amplified, and emitted by the loudspeaker, and then it comes into the microphone again and so on, and becomes a heterodyne whistle. The note of the whistle depends on certain characteristics, acoustic and electrical, of the medium and the circuit, and I could, by varying these, get a number of different notes, and indeed I could arrange that it should vary systematically. In fact, if we let Longuet-Higgins loose with one of his computers, it wouldn't be long before he produced a program which as soon as any noise came in to set up a heterodyne whistle, would produce just those notes which are the opening bars of, say, one of the Brandenburg Concertos. This is an example of how the music of the biosphere (one of Monod's favourite metaphors – see e.g. *Chance and Necessity*, p. 114) could be extracted from the random noise of the fluctuations and mutations which he rightly sees as the basis of the evolutionary process.

This is not a form of explanation peculiar to the biological sciences. The chemists and the physicists also use it. There is a theory of the evolution of the elements. If you want them to explain how the elements come about, they draw attention to the fact that, thanks to quantum indeterminacy, there will

124

always be fluctuations in energy levels and that even very difficult energy walls can occasionally, not so much be surmounted, as burrowed through; and so in the fullness of time all sorts of different configurations will be tried out. It is a question of quantum mechanics — what configurations are relatively stable? — and it will be found that in the fullness of time hydrogen sometimes gives rise to deuterium or helium. One can thus build up an account of how the elements have evolved, which depends essentially on the overall structure of the quantum levels of the nuclear particles. In the same way, many chemical explanations — that of the precipitation of silver chloride in terms of solubility-products — pay attention not to a genetic account but rather some overall properties. This type of explanation is a rational, rather than a genetic or a regularity explanation. Moreover, it is open-ended rather than sewn-up, because stability is always relative to context. This last point is the most obvious one when we are considering the biological sciences, where the reason why a species has survived is because it is well adapted to its habitat, and has been able to fit well the ecological niche which is available to it, and adapt itself to any others that may be open to it.

We can see how, over the centuries, the endless interminable fluctuations of DNA molecules and proteins will ensure that every possible combination is likely to be tried and subjected to ecological constraints, and any viable one is likely to be explored and realised. When we come to consider man, we want to say two things. First, that if the question: 'How has man evolved?' is asked in a purely genetic way seeking for a purely genetic explanation, we agree with Darwin and Huxley. But if we ask the other question: 'Why has evolution produced something like men?', then we shall pick out those features of *homo sapiens* which have made him particularly able to survive, namely his ability to anticipate, to accommodate to, and often to control, his environment, and his ability to concert his efforts with other members of the species; or, to put it another way, the fact that he has a mind and can talk. That is to say, if we ask the question: 'Why has man evolved?', we shall find ourself talking about the fact that man is endowed with a mind and can communicate with other men by means of messages; those very excellences wherein man is like the angels.

The Darwinian controversy produced a great sense of shock — it was thought to refute the standard arguments for the existence of God, the Argument from Design; and so in a sense, it did. But before we are too sad about the fact that we no longer have any need, as working biologists or working scientists generally, for the theistic hypothesis, we should look rather carefully at the sort of God whose existence can no longer be proved. Yesterday, I protested at the view Longuet-Higgins was taking when he went 'shopping around' for gods.

But today while still maintaining that there is a certain logical as well as theological inappropriateness about the consumer approach, let me go with Christopher and Tony on a shopping expedition, as Tony tries to find a god that Christopher can believe in. Since they have certain left-wing leanings, we will borrow the metaphors from the means of production. The first god that Tony offers Christopher, is a good eighteenth century working man, the great craftsman, the Demiurge who is able to make the world as it is, because this is exactly how he designed it. And instead of taking up Longuet-Higgins's objection that this is an out-of-date conception that has been rendered obsolete by more recent technological developments, let us just look at it in the spirit of *Which?*, noting its merits and demerits. It has certain merits; not only did it explain the biological phenomena and some of the astronomical ones, but also it gave us a doctrine of particular providence, which explained why I am here and what I ought to do. It gave some sort of meaning to life, and this should be given two or three ticks in the credit column. But it had certain bad points; in particular, it had no room for any sort of human freedom or spontaneity – the metaphor was that of the potter's vessel; and, with due respect to St Paul, this ought not to be the dominant metaphor of any theistic understanding of the nature of man. We are not playthings. If we are to understand ourselves at all, we must see ourselves in some sense as being in the image of God, in the likeness of the ultimate reality, and therefore must ascribe to ourselves some sort of freedom. Therefore, quite apart from any modern objections, we can see that there are certain moral and theological objections to the first god that Kenny offered Longuet-Higgins.

We turn to a more modern version; not a craftsman, but a manufacturer engaged on mass-production. We now see God as the great Lord Nuffield, who produces innumerable copies of the same standard design. This is the god of the classical chemist, who created enormous numbers of atoms that are exact replicas of each other. This view of the Demiurge producing a universe on a certain standard pattern has a very respectable ancestry. But as we begin to look at it, even Lord Nuffield finds himself redundant. We begin to play down the Demiurge and play up the patterns – the Forms. We consider no longer the god of the great chemist, but the god of the chemist-turned-mathematician, where what we stress are the underlying configurations of possible states of affairs. We want to see what paradigms are possible, and we pick out those patterns and look on them as our explanation of why the cosmos is as it is. And then we move a stage further, and try and see a more 'Waddingtonian' god, with craggy, chreodic features in which we are no longer emphasising only law-likeness, the intellectual elegance to which Longuet-Higgins paid eloquent tribute last night, but also the sense of potential-

126

ity – the sense that there is always a bit more to it, that the universe is in some sense more creative, it is always finding out new things. When we consider the biological phenomena of the world, we are impressed in two different ways: one is the immense prodigality of the biological world; the other is the inexhaustible variety and the endless initiative of the biological world. I shall not discuss the former, as I do not think we can extract a clear moral from it; but the latter we can see as giving us a clue to another possible god for Longuet-Higgins to believe in. Even so, he may not be fully content: this god, he may feel, still does not explain the whole range of human affairs and, in particular, neither answers our moral needs nor gives an entirely satisfactory account of the existence of mind. Certainly Monod feels dissatisfied. Not only will his dissatisfaction aid our thinking about the nature of mind; but so too will our dissatisfaction with some aspects of his account reveal one important requirement of any overall description of what the world must be like if it is to describe adequately a world in which minds exist.

Monod writes under a great sense of shock – he is appalled at the idea of 'Nature red in tooth and claw'; he has the same reaction to it as the Victorians – 'so careful of the type she seems, so careless of the single life'. He keeps emphasising how nature seems to be entirely indifferent to our purposes and deaf to our music. It is a real feeling that many of us have. But we should see it as the other face of freedom. If we are autonomous agents capable of choosing a number of different things, it is necessary that the Nature in which we make these choices shall be indifferent to our choices. In many ways of course, we are not totally free. I am not effectively free to eat arsenic; if I eat arsenic, my life would come to a rapid and painful end. And when we say that man is free, what we have in mind is at least two things. First, that we are able to think and feel any of a wide range of things without suffering deleterious consequences; and, second, there is a reasonably wide range of things that we are able to do without suffering any deleterious consequences. And this is the reason why we have the sense of aimlessness and loneliness that Monod puts forward when he tries to see the world: 'If man accepts this message in its full significance, he must at last wake out of his millenary dream and discover his total solitude, his fundamental isolation. He must realise that like a gypsy he lives on the boundary of an alien world; a world that is deaf to his music, and as indifferent to his hopes as it is to his suffering or his crimes' (p. 160). Very eloquent, but not new. We didn't need Monod to tell us that here we have no abiding city, that we are pilgrims and sojourners as all our fathers were; and it is not because of any discoveries of molecular physics that we can come to see this, but because of our first consciousness of what it is to be a person, our first consciousness of being number one, a self that is aware of itself as an autonomous agent. That is to say, if I have any sort of freedom, then necessarily I must see myself as being something separate from the world. It is part of the logic of the first person singular that I should see myself as:

'I, a stranger, and afraid,
In a world I never made.'

The Old Covenant, which Monod says is broken, is set over against this as something coming before our awareness of our own autonomy; and for most of us most of the time it is possible to live under habits of mind that we have inherited from our parents and a number of conventional wisdoms and traditions we have taken over from our own society. It is only occasionally that we come to realise that all things are possible, and that although conventional wisdom lays it down that this, and this, and this, is the way of doing it, nevertheless, it is possible to think of doing other things, and even to do them. And it is only then that we become aware of ourselves, or as the myth of the Garden of Eden puts it, that we are naked. It is the nakedness that I want to bring out at the moment, which comes to people that are conscious of their own autonomy, as they realise that they have no vestiges of customary costume to clothe their own inadequacies. The Old Covenant was a covenant of servitude; but we have a very ambiguous attitude towards emancipation, and emancipation is a painful process whether for Adam, St Paul, or Ivan Karamazov. It is difficult to come to terms with being able to make free choices, and therefore one is bound to look back on the simple certitudes of one's youth with nostalgic eyes and to think that because we were at home and everything was happy then, we are necessarily orphans now. But this conclusion doesn't follow, any more than the conclusion which Monod actually drew from the discoveries of molecular physics.

Not only does Monod's conclusion not follow, but one crucial aspect of his account is itself inherently unsatisfactory. (In case you feel that I'm being a bit unfair in concentrating on one aspect of one man's world view, I should explain that I have a reason for concentrating on particular examples. I'm not sure that I can give a complete overall account of the world or how we should go about constructing one. But what I think I can show is a way — a schema of argument — in which a range of different alternative accounts can be refuted. And from this we can obtain some guidance about how we ourselves should go about the job. See also p. 149). Monod is obviously uncomfortable when he talks about the Kingdom, and about Values and Knowledge, because he is committed to a certain doctrine of the objectivity of knowledge which runs counter to another doctrine he has about all values being essentially non-objective, and so he has to have a certain rather implausible appeal to some fundamental ethical axiom, or an arbitral choice, which will justify our having any concern for knowledge at all. And this can't work; that is to say, it must be incoherent to separate too sharply knowledge from values, at least so long as we believe that knowledge is worth pursuing and that truth ought to be believed. Any definition of truth or any criterion of objectivity which cannot accommodate this fact must be rejected because it is denying its own warrant of acceptability. You can't have a theory of knowledge which makes knowledge not worth pursuing or an account of truth which leaves it, as it were, an open choice whether you believe what is true or not. To do this, would be to cut off the branch on which you are sitting. It was an intimation of this that led me to formulate my Gödelian argument last year, and to claim that a whole range of otherwise possible overall philosophies must be in some sense self-defeating.

Although knowledge is in some important sense objective, and although values must, if they are to be authentic, be autonomously adopted, it doesn't follow that it is merely a subjective whim whether one should choose to desire knowledge or not. It is never enough, therefore, to say merely that knowledge is objective; we need also to recognise that it must be personal, that knowledge is necessarily concerned with the knower as well as the known. And so, if as autonomous agents we aspire to affirm values that are objectively valid, we shall have to base them on certain fundamental facts of our existence. Not only are we independent of the world and to that extent set in a world that is indifferent to us, but we are in a world which has a matter of fact evolved us rational intelligent beings, and which, we now see, evolved us in a sense naturally and as a matter of course. And if we embark on any sort of metaphysics, we shall find it difficult to ascribe to the cosmic reality any lesser reality or any lesser rationality than we ourselves possess. When yesterday Longuet-Higgins was talking about cosmic purposes, Kenny began to make this point against him, that unless he was prepared to talk about something which could probably be described as God, his position was going to be necessarily incoherent. And I should maintain, quite generally, that any view of the universe large enough to include the fact that minds exist will itself have to be so large, so rational and so personal as to deserve the appellation of 'God'.

To sum up: Monod is wrong and the French Existentialists are wrong and many sceptical philosophers are wrong, all for very similar reasons; they err both in their diagnosis of what Monsieur Monod calls the 'soul's sickness', and in the remedy that they offer. The breakdown of the Old Covenant is real; but it is due not to our discovering that values aren't objective but to our realising that we are free, enfranchised with the knowledge of good and evil, and able to choose the evil rather than the good. Values indeed are empty unless they are adopted by our own free choice, and we do not have to choose to adopt them. But from this it does not follow that if we choose them, they are just vanity, or that to choose them is an *acte gratuit* or inherently absurd. It doesn't follow; but it may seem to. On the Christian view, it is part of the human predicament that although man needs — and once again I quote — 'an ideal transcending the individual self to the point even of justifying self-sacrifice' (p. 165), he cannot, as an autonomous self, establish one from his own resources. For that, something else is needed, something from outside involving not the austere concepts of morality and autonomy, but the more intimate concerns of personal relationships and love. So far as natural theology is concerned, the delineation of these must remain for ever in the optative mood. And to go any further, and to argue that these *desiderata* are, fortunately for us, in fact fulfilled, would involve appeals to Revelation which lie beyond the scope of the Gifford Lectures.

Discussion
LONGUET-HIGGINS
I'm sure we would all agree with much of what John has said; I certainly do. I should like, though, to try and restore the balance a little bit by putting right what are, in my opinion, slight distortions of Monod's position which I think John has made.

As I understand Monod's position, he starts from a certain intellectual problem which is that although we have no reason to suppose that there is any rhyme or reason about what atom is going to hit what molecule next, or what cosmic ray is going to hit what DNA molecule in what place, none the less, evolution does result in the appearance of living creatures and forms which show this characteristic of teleonomy. Now, his provisional answer to that problem, which he develops at some length with characteristic elegance — its really a modern development of the Darwinian theory — is that we have at work a selective process which is particularly effective because of the way in which, once you get started on a particular teleonomic system, the system can bootstrap itself and raise itself by its own shoe-laces, by incorporating new and biologically viable gimmicks which will help it to explore more thoroughly, to make the most use of its present ecological niche. As he puts it (in my own translation):

> 'The extraordinary stability of certain species, the millions of years which evolution has covered, the invariance of the fundamental chemical plan for the cell, can only be explained by the extreme coherence of the teleonomic system, which in evolution has thus played the role both of guide and critic, and has preserved, amplified and integrated only a tiny fraction of the possibilities offered to it in astronomical numbers by the roulette wheel of nature'.

That is the necessity which he talks about. When Monod speaks of necessity, he's talking about the inexorable laws which determine that if you bring two DNA molecules together which don't fit, then they won't fit; but if you have a hereditary mechanism of the sort which he describes in such detail, then this will act as a very effective stabilising influence on any advantageous changes which occur. So I think that John has slightly misrepresented the precision of Monod's thought about chance and necessity. That's my first point. (I might add, incidentally, that Monod does say on the cover of his book: 'It is inadvisable today for a man of science to use the word "philosophy" in the title or even the sub-title of a book. To do so is a guarantee that he will be greeted with contempt by scientists and at best with condescension by philosophers.')

I'd like to take up some of the later points in John's paper. I think that he has correctly identified the weakness of the principle of objectivity as an ethical principle, if one accepts, as one reasonably might, a sharp distinction between knowledge and value. But I think I should just say what the principle of objectivity actually is, because Monod in fact defines it quite precisely early in his book where he says: 'The corner stone of the scientific method is the postulate of the objectivity of nature, that is to say, the systematic refusal to entertain as a true explanation of phenomena any interpretation in terms of final causes or plans'. In other words, what Monod is rejecting in adopting the postulate of the objectivity of nature is exactly what John calls the doctrine of particular providence. John's statement of the doctrine of particular providence is that according to the doctrine everything that happens is foreseen and is the intentional handiwork of the Creator, so for John to reject that doctrine is for him to embrace Monod's principle of objectivity.

So let's see where we go from here. John Lucas and Jacques Monod seem to agree in essence as to how the phenomena of evolution should be interpreted. The real point of disagreement comes at the end, when we have the problem of how does one fit together one's knowledge and one's ethics. Here I should like to start a hare. I'm sure that many people in this room will be familiar with some current trends in the subject of artificial intelligence; it turns out to be very illuminating, and helpful, to represent knowledge as essentially procedural. Let me explain. To know that this is a glass which I could drink out of, is to have at my disposal a certain routine which I can go through when drinking; that is the cash value of my knowledge that this is a glass and that the stuff in it is water. Now, if ethics isn't about what we can do and about the decisions we can make, then I don't know what it's about; and I think that if you look at knowledge as essentially a useful set of tips or rules of thumb, you see that the dividing line is likely to become slightly imprecise between a system of knowledge and an ethical system. I think that this might be a way of interpreting what Jacques Monod means when he says that to accept the principle of objectivity is to adopt the basic proposition of an ethic, the ethic of knowledge; and I believe that that remark has to be thought about for a very long time before one can evaluate it properly.

LUCAS

I will just make three points. I don't want to argue about Monod's scientific doctrines; it is outside my competence. Of course, I was trying to say the same as he, but to put a different complexion on it. He talks about the chances of the various configurations. I say in one sense this is random, but then I point out another way in which it is only to be expected — it is rational; and so when he says 'chance versus necessity', or 'chance versus rationality' as I would rather put it, I want to concede that there is something chancy about it, but there's another type of reasoning which shows it nevertheless, not to be irrational, but on the contrary rational. Secondly, I agree with Christopher that I am rejecting the doctrine of particular providence; not just simply because it's been shown to be wrong scientifically, but on moral and human grounds because it denies human freedom. I tried to show that if one is going to have any account of man which allows man to be free, then there is bound to be some sort of randomness in the universe just following from the fact that, in one sense, man is free.

The third point which Christopher said was true, was that we should meditate on the implications of Monod's fundamental ethical postulate. Indeed I think we should; but I think that, if one tries to think it through, it then leads to the rather large metaphysics that I espouse, rather than the thin and arbitrarily adopted one that he is offering us.

WADDINGTON

At one point John said that man must see himself as the image of ultimate reality; then he went on to speak of man as being like a gypsy existing in precarious camps in the margins of a world he did not make. I think this second notion, of the alienation of man, arises from the very great and, I think, somewhat mistaken emphasis which John puts on the notion of personal freedom.

It seems to me that this great valuation of freedom is primarily a Jewish and Christian attitude. I do not think it is a Chinese or Mohammedan attitude, and in fact it is not a necessary attitude at all. In one of the lectures last year I mentioned the attitude which holds that the really and truly good man is one who does not have to exercise his personal freedom to do good, but who just naturally behaves in a good way without thinking about it. This surely is the ideal of a man, who is so much a part of nature, so much 'the image of ultimate reality', that he simply acts in a way consonant with that reality. The whole notion of an objective world independent of ourselves, on which Jacques Monod lays so much stress, and the notion of an ethical postulate that that truth is something which should be believed in, seems to me a clumsy and misleading way of saying that man should try to mould himself into conformity with the underlying structure of the universe. Really this applies to all living things throughout the course of evolution. I think man has already evolved to a state in which he is in much closer conformity with the structure of the universe than, say, a snake or a fish, because he incorporates into his knowledge an enormous amount of that structure which they are totally unaware of. So I think I would lay the emphasis not so much on man's freedom to behave as he wills, regardless of the nature of the universe, but rather on his still inadequate success in becoming fully tailored to fit into the world in which he is living; and that the more he fits it, the less need he will feel for 'freedom' in the sense of disconnection.

KENNY

I'd like to take up two points that John made — the first, the point about cosmic loneliness and the second, the point about particular providence.

It seems to me that both Monod and Lucas are foisting on us as part of the real human predicament a totally imaginary predicament which is that of cosmic loneliness. They say that if there is not a purpose in the cosmos, if the non-human part of the cosmos does not have purposes and things don't happen for reasons, then we should be lonely, we should be like gypsies living in an alien world. This seems to me totally wrong — each of us has quite enough, more than enough, other human beings whom we are constantly living with to keep us from loneliness if we make friends with each other. It would no doubt be very comforting if we could be assured that the universe was controlled by an intelligence that was powerful and benevolent to human beings. But if the universe was controlled by a powerful and hostile intelligence we would be equally free from loneliness but this would not be at all a comforting thought.

The solitariness of man in the universe is, I think, just an imaginary trick which is played on us by Monod and Lucas, and essentially it is based on the anthropomorphism of thinking of the rest of the universe as if it was a human being but a human being who wouldn't talk to us. This is brought out particularly well by Monod's reference to gypsies. After all, what is wrong with being a gypsy? The thing that makes being a gypsy uncomfortable is the existence of rural district councillors who keep on wanting you to move on, or the existence of Nazis who are going to put you into a gas chamber. If you were gypsies who had the world to yourself, this would not be a lonely but a happy state of affairs; you would be kept from loneliness by the other gypsies and you would be free from

133

persecution by the non-gypsies.

I now turn to the second point, about particular providence. I think that John was wrong to say that the doctrine of particular providence had been refuted by science, if the doctrine of particular providence is the doctrine that the life of each individual is the result of an intentional decision of God. I think that, at most, what has been refuted is the idea that the life of the individual is the result of a number of intentional interventions by God in the course of his history. But there are at least two other versions of the doctrine of particular providence which I do not see to have been refuted at all.

The first was the idea that while the life of each individual, say the life of John Lucas, is part of a divine plan, it was part of a total divine plan made by God before the world began so that out of the many possible worlds which he could have created, he created one in which the operation of natural laws, in which he did not intervene, would produce John Lucas with the history which he has. John may well say that this would involve determinism, which for other reasons he is determined to reject.

But there is also a version of particular providence which is not deterministic: it is simply that it would be possible for God to leave John free in his choices but in virtue of his power and knowledge, which so much surpasses even that of John, to adopt a strategy which he foresaw, no matter what choices John made, would give him a certain history. This would, of course, not mean that each particular action of John's was the result of providence but that John's overall life was.

John, as I said, rejects determinism because he thinks that it is incompatible with autonomy. I argued last year, that this has not been shown; but I want to conclude by exposing an equivocation which John made use of in his attempt to show that it was not only a truth but a logical truth that anybody who was free must be cosmically lonely. He said: 'If I have freedom, I have to see myself as a stranger and alone because nature has to be indifferent if we are to be free'. Here he was using a blatant equivocation on the word 'indifferent' which may mean either non-determining or uncaring. John believes, wrongly I think, that if we are to be free nature must be indifferent in the sense of being non-deterministic; but even if he's right in that, it doesn't follow that nature must be uncaring if we are to be free. Many theologians have believed in a determinist* God while believing that far from being indifferent to what his creatures did, he cared enough about it either to reward them with heaven or punish them with hell in consequence of it.

LUCAS

I'm saying it is not with a determinist God, but with a

---

* I should, of course, have said 'indeterminist'!

non-determinist God where issues of indifference rise. More generally, I think it would be perfectly fair to go very carefully over the use of the word 'free' which, like 'necessary', 'reasonable', 'chance', and 'random', is systematically ambiguous. I don't think I made the mistake that Tony thinks I have, but one wants to assure oneself of this independently of anything that I say. I didn't say that the doctrine of particular providence had been refuted. What I said was that one traditional argument for a version of this doctrine had been shown to be unnecessary; that is why I used the metaphor of redundancy which I first of all applied to the working man, the Humean artificer, and later to Lord Nuffield. But I also say that the doctrine under consideration was one whose passing we should not regret, because it was open not merely to logical, philosophical and scientific, objections, but to moral and theological objections too; and this comes out in the way Tony tries to gloss his position by smuggling in, as a special case of doctrine of particular providence, a doctrine of general providence, where he sees the deity as not determining each particular choice that a man may make, but only adopting an overall strategy which will, one way or another, produce certain rather generally specified results. This of course may very well be true. What actually happens with the pin-table (p. 10) is that there is a certain general overall strategy which, without determining the exact path of each particular ball, nevertheless, produces a general overall result; and it would be perfectly possible — in fact, Tony is probably right in suggesting that it is possible — in some ways at least to see a certain grain in the way that things happen; which is why we have ideas of poetic justice and the idea of certain values naturally working themselves out in the course of human life. This is stuff about which novelists are much better able to talk than I, and equally I shall leave it not so much to the novelists as to the poets and the prophets to persuade you that it is not just simply a cosy little dialogue of Jacques Monod and myself, each telling the other how lonely he feels, but a very common feeling. True, it doesn't logically follow that I've got to feel lonely; no doubt, many people do not have any sense of being separate from the whole of reality — and Wad approves. Nevertheless, loneliness is something which is, at least potentially, wrapped up in the concept of the first person singular. Not only is it logically necessary, but it is part of the elucidation of the concept of 'I', that I should have occasion to say, in the words of A. E. Housman:

> I, a stranger, and afraid,
> In a world I never made.

Now Wad, quite properly, thought that I was speaking rather ambiguously about freedom, and that this was a Hebrew notion which we don't find in all other cultures. This is true, and I shall leave it to be pressed a bit further tomorrow. At the moment I only want to say 'Yes, there is an ambiguity', and that if we have any notion of autonomy we find two further strains: one, an intense awareness both of oneself as an independent ego and of the egocentric predicament, which is what we should now see as the origin of sin — a sense of alienation, of being cut off from everything else; the other, a desire not to be cut off, a desire to merge one's purposes with that of the cosmos as a whole, as we heard yesterday, as well as, of course, joining up with other gypsies in our earthly pilgrimage. Both these, I think, are real intimations of our situation, both seem to be compatible; although as I have hinted, both can be reconciled only in a much wider world view.

Tenth Lecture. Questions and Answers

KENNY. This lecture will be a question-and-answer session among ourselves, and I'd like to begin by explaining the batting order. You will remember that we've been, some of us, rather shy about offering a definition of mind, but I think there have emerged four different criteria for distinguishing minds from non-minds which could be put in order like this:

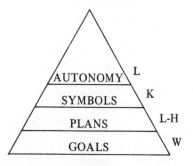

For Wad, goals are perhaps the most important thing separating minds from non-minds; for Christopher, it is plans and procedures; for myself, it is symbols; and for John, it is autonomy. You can arrange these in a pyramid, because the lower down ones apply to more entities than the upper ones do: there are very many things that have goals, but very few have John's particular sublime autonomy. In our discussion we shall start at the bottom and work upwards, taking turns to answer questions in the following order: Waddington, Longuet-Higgins, Kenny, Lucas. So it is Wad to whom I put my first question.

You referred in your last talk, as a possible explanation of the emergence of goals in nature, to recent work of Kauffman. If I understood it rightly, this work shows that a series of random switching instructions given to an array of lights may give rise in quite a surprising way to an orderly pattern of flashing. This work, it seems to me, may show how from an anarchic beginning we can get the development of a cycle, but the existence of a cycle isn't the same thing as the existence of a goal. The things that occur in a cycle don't necessarily occur for the sake of the cycle, or for the sake of one stage of the cycle, as you can see by considering say the oxygen cycle. It's only where there is a cycle in which things occur for the sake of a stage in the cycle that there are goals, I think. Now it isn't

137

at all easy to identify a stage of a cycle as a goal. Bertrand Russell once tried to do so by saying that the goal of a behaviour cycle of an organism was that stage of the cycle which brought the activity to an end, the stage of a cycle which was followed by quiescence. And of course, that won't do as you can see by considering the well-known phenomenon of the pyjama cycle. It would follow from Russell's account that all human activity is directed to the goal of getting into pyjamas, because the one human activity that is most often followed by periods of quiescence is the donning of pyjamas. So I'd like to ask you, Wad, how you identify goals, and how you distinguish between cycles and goals?

LONGUET-HIGGINS. Wad, I gather that you regard the possession of goals as central to having a mind, and that you offer the chreod as a concept useful for thinking about goals. Do you regard the existence of chreods as an explanation of goal-directed phenomena, or as a fact requiring explanation?

LUCAS. Are goals, as you understand them, necessarily accompanied by intentions, or not?

WADDINGTON

In reply to Tony Kenny I did not mean to imply that Kauffman's work was an example of a goal appearing from an anarchic background. Kauffman observed that the system tended to go into a limit cycle; it got into a state in which it was going round and round one repeating sequence of changes, and if it was disturbed it tended to get back on to this limit cycle or on to one of a small number of alternative cycles. This situation is rather similar to the phenomenon known as homeostasis, in which a system gets into a stationary condition, to which it tends to return if disturbed. The concept of a goal is, I think, only applicable to systems which are continuing to change, and which tend to change along a defined course even after disturbance. These are systems which show homeorhesis rather than homeostasis. The point I wanted to make was only that the appearance of a limit cycle, like homeostasis, represents a rather elementary type of order, which can come into being in a system which appears at first sight to be quite chaotic. The appearance of a chreodic system, which one could speak of as having a goal, would, of course, be a more complex form of order.

Now to turn to Christopher's point. I certainly regard chreods as things to be explained and not as themselves explanations. It is a concept which I think is useful to keep in mind when we are dealing with the higher levels of Tony's pyramid. The point of it is that it reminds us that in discussing mental events we are probably dealing with systems very unlike the ordinary causal systems we usually come across. In these the nature of the output is usually closely connected with the nature of the input, and a change in input leads to a change in output. In a chreodic system, however, the output

138

may be almost independent of the input provided only that the input falls within a certain range; but if it falls outside that range, the output may switch over into some quite different type. I have indicated the nature of such systems by the model of valleys with watersheds between them, but this is, of course, merely a descriptive device. It does not in any way explain why the system should behave like this.

Now there is John's question. Is a chreod or goal the same thing as an intention? No, of course it is not. The idea of an intention implies consciously formulating a wish to attain some particular goal. I am discussing the basement level of Tony's pyramid, the biological processes occurring in systems which are prior in evolution to the appearance of man as a language-using animal. I am looking in this pre-human world to try to see the foundations out of which such things as intentions may have evolved.

Questions to Longuet-Higgins

KENNY. I wanted to ask a question arising partly from last year, when you said, Christopher, that the difficulties which John had presented against artificial intelligence from the findings of Gödel and Tarski could be avoided if one switched from the indicative use of language to the imperative use of language so that one could avoid the difficult concept of truth. Now, it seems to me that the important thing is not what makes a difference between the indicative and the imperative (which is a difference between two sorts of speech act, assertion versus command), but rather what is in common to the two, namely, the notion of matching between the world and a bit of language. In the case of a statement, the onus of match is on the statement to match the world, i.e. to be true. In the case of a command, the onus of match is on the thing commanded, to match the command, i.e. to be obedient. Now it is really the notion of match that gives rise to the problems which troubled Tarski. And I wanted to ask you whether you think that your concept of meaning, as essentially the procedures which the words or sentences used give rise to, can avoid the difficulties of the concept of matching.

LUCAS. I, too, want to sit on Longuet-Higgins on his being too much wedded to programs. I think there are in fact two people here: there is Christopher, a very reasonable person who is extremely intuitive, and often thinks quickly without being programmed, and also there is the Professor who is very clearly wired up in the imperative mood and lays down sharp instructions about what is to happen. And my point isn't so much Tony's that instructions often raise the question whether they are going to be obeyed or not, but rather the point that I tried to make last year that even within the formal confines of mathematics, not every problem can be algorith-

139

mically solved. And I still want to press Christopher whether he does or does not believe that every program, that every rational procedure must be algorithmic?

WADDINGTON. I want to ask Christopher a question about consciousness. We all know something about our own consciousness because we experience it. We know also that it can occur in a number of different forms, for instance under the influence of alcohol or other drugs, and people assure us that it can be altered by various techniques of yoga, controlled breathing and so on. Moreover, if animals such as dogs and cats are conscious — personally I don't see how we can ever tell whether they are or not — if they are, surely their consciousness must be of quite a different character to our own, since for instance they rely on different modalities of sense, such as smell. We have, therefore, I think to accept the possibility that there may be different forms of consciousness possible in the universe, so the question I want to ask is: can we suppose there is anything corresponding to consciousness in a computer? Can Christopher think of any way of deciding whether one of his intelligent computers is conscious or not?

LONGUET-HIGGINS

First of all, Tony's question about the lecture last year in which I tried to develop an imperative theory of meaning. I did so by contrasting indicative languages, such as the first-order predicate calculus, with imperative languages, such as programming languages, in which you write your stuff and feed it into the computer and something actually happens. And I was adopting the imperative language as a way into the theory of meaning because it seemed to me that what we were really trying to understand was the way language works, and the way it's used. I hadn't very much to say about the way language is produced, but I thought there was something worth saying about what happens to utterances when they enter the hearer's ears. I was trying to suggest that the meaning of an utterance in a natural language is possibly to be compared with the meaning of an utterance in a programming language, because an utterance must mean something *to* somebody and one must never forget the person to whom it does or does not make sense.

The idea was that the meaning of an utterance is to be identified with the process of thought which it represents. The point of a programming language for a computing system is that you can specify in that language a very large number of processes of interest and usefulness — to us, of course, rather than to the computer, but that's by the way. Perhaps a better way of making the point would be not to talk about the indicative or the imperative but the infinitive. Let me try to illustrate this idea with an example. I will write on the blackboard a simple mathematical expression, namely $(2+3)\times5$. I suggest that what that means is: 'to add 2 and 3 and then multiply the result by 5'. The unadorned infinitive clearly specifies a procedure. So far there is no 'speech act', but I can turn this infinitive into a speech act simply by putting the words 'I order you' in front, and then it becomes an imperative. So one might suggest that sentences in natural language are ultimately built from infinitives, to which one has to append a proforma in order to turn the thing into a statement, command or question. One must of course distinguish the *procedure* of adding 2 and 3 and multiplying the result by 5, from the answer which results from this procedure — every logician knows the problems that arise if you don't.

I hope this may help to answer Tony's direct question whether procedural semantics avoids any of the difficulties associated with the matching of statements to states of the world. In one way, I think, it does. To learn a language is to discover how to modify one's views and habits on the basis of what people say, and conversely, how to influence other people's world views and dispositions by choosing one's words appropriately. If utterances are regarded as functions from states of the hearer's mind to other states of his mind, then once the hearer has learned the language, the only matching problem that remains is that of judging whether the world as he knows it agrees with the view of it which is suggested to him by any given declarative sentence. This is my own view of the problem of establishing the truth of a statement in natural language, and it fits in with the view that the truth of a statement cannot be ascertained in principle till we know what it 'means'.

To John — how do I cope with the non-algorithmic aspects of reason, and I am wedded to the view that reason must be algorithmic? Well, I can't really form a coherent concept of reason or rationality except in terms which will make it clear to me how one's thoughts follow one another, or how one's decisions are related to the situations in which one finds oneself, or how one's beliefs are determined by what one is told, or how moral considerations enter into the decision to do one thing rather than another. I think this is simply the way we think and the way we discuss matters. Now, of course, it might very well be the case — and it almost certainly is the case — that real human beings are not deterministic organisms for which we could give a total specification. But this issue of physical determinism versus indeterminacy is really utterly unreal as an issue of practical or philosophical importance. I mean, we have the Indeterminacy Principle anyway, we have the utter impossibility in practice of giving a description to the state of motion of all the atoms and molecules in one's brain and one's environment, so even if our brains *were* causally determined in the strict sense, it wouldn't really make any difference to the way we could usefully talk about the relation between physics and what we do. So I'm not deeply concerned about that issue. I'm not wedded to the idea that we have to be deterministic organisms in order to be algorithmically describable. On the contrary, I regard the chreodic character of the human mind as quite astonishing. With all the perturbations which enter one's consciousness, the fact that one can stick to a train of thought, even roughly, is a very remarkable fact, so it seems as if we are to a very considerable extent helped by nature to behave like automata for the purpose of thinking. But supposing we were not in fact describable as automata, and that there was an element of randomness — dare I use the word? — which meant that no account whatever could be given of why I did a certain thing; I don't feel that that fact ought to cause any more trouble to my view of the mind than to John's view. I can't allow that John would be able to say: 'He did it because he wanted a piece of chocolate more than he wanted a piece of toffee', while I couldn't make some corresponding algorithmic statement of what went through my mind; if I couldn't, then John would have to withdraw his description too.

Finally, I have to answer Wad's question — what would make me regard a computer program as conscious. This is very hard; I find it much the hardest of the three questions, I must admit. One can make a few obvious remarks about consciousness, such as that a sufficient condition of consciousness at a given time is a person's capacity to remember what happened at that time. If I say to you: 'What were you doing at 3 a.m.?' you probably won't be able to give me the answer because you were asleep, you were unconscious. On the other hand, if you tell me you met somebody for lunch and he said this and that to you, that seems to be compelling evidence that you were conscious at lunch time. So one might be tempted, and I once was, to define consciousness as the rate at which your memory changes — when you are asleep your memory isn't being modified by experience — but that doesn't seem to be quite good enough. Because one can certainly be conscious — one can be sure that one was conscious at a certain time — and be unable to remember or report anything about one's state of mind at that particular time.

I think that another basic necessity in defining consciousness is the concept of a world model. A conscious being has to have at least some internal representation of the world (that's a necessary, not a sufficient condition for being conscious). Now, it strikes me that a possible way of defining consciousness — at least in animals and in human beings — is to say that one's degree of awareness or consciousness is essentially measured by the rate at which one is currently reorganising one's world model. This definition is just sufficiently blurred at the edges to leave open the interesting question of whether we are conscious when we are dreaming. Because when one is dreaming one is in a sense reorganising one's world model, and indeed we can remember our dreams. It also allows that we can be conscious without receiving sensory input from the world; and when one is deep in thought but blind to the world, surely one is still conscious. So I would say that the rate at which one is reorganising one's world model is really what we have in mind when we talk about one's degree of consciousness.

Now, the final question of course, is what about computing systems? How can we tell whether a computing system is reorganising its internal representation of the world, and how fast? We can only know, of course, by studying the program which has been written for it. Now, I'm not one of those who think that the fact that a human being wrote the program is philosophically frightfully important in our discussion of the nature of the activity which goes on inside the computing system when the program is actually running, any more than Tony Kenny is worried by the thought that the man from IBM might appear and say to him: 'Very sorry, sir, you have to stop talking now — I've come to service you'. I regard my consciousness as my own property, whether or not I have been produced by the intentions of somebody else. So I think we'd simply have to look and see whether the program was reorganising in some interesting way its world model, and if it were, then according to my definition, it would be conscious at the time.

WADDINGTON

But Christopher, surely you could write a program which would cause your computer to be continually revising its world model, either in a random fashion, or in response to various external inputs, so that according to your definition you could easily arrange for your computer to be conscious. I don't think I am convinced.

LONGUET-HIGGINS

Well, I'm not entirely happy with my definition; but I've tried, I've done my best!

Questions to Kenny

WADDINGTON. I think Tony has been the one amongst us who has laid most stress on logical arguments. For instance, he argued that language could not have evolved, because evolution implies inheritance from ancestors, and clearly the first users of language could not have learnt this from their ancestors, who by definition would not have language. This is an argument based on one type of logic. Now there are several types of logic available, and I want to ask Tony what he thinks is the status of logical argument. Are the different types of logic simply different tools which we can pick up and use when appropriate, as we use a plough when ploughing a field, but use a different machine, a reaper, when harvesting the crop? Is any particular kind of logic just a tool suited for some tasks and not for others? And if we use a logic and get what seems to be the wrong answer, which does not agree with what we had previously thought, under what circumstances do we accept the logic and say that we must have been wrong previously; or do we stick to our previous opinion and reject the logic as being unsuitable for use in this connection?

LONGUET-HIGGINS. A few lectures ago, you defined mind as a capacity for intelligent activity, and intelligent activity you defined as activity requiring the use of symbols. You also implied that language is essential to mind. Would you agree that it is possible to think symbolically without necessarily being able to communicate in language?

LUCAS. I want also to try and tie Tony down a bit about language and how far this presupposes mind. I still feel, Tony, that you are too wedded to language and are always explaining mind in terms of language, and while I don't want to deny the importance of language to mind, it seems to me that language only can be understood and meant in the context of one mind communicating to another; I can never make out what you really think about this.

KENNY.

Firstly, I'd like to say to Wad that I didn't argue that an evolutionary explanation of the origin of language was impossible; I tried to show that there was an important difficulty which I thought had been neglected. But Wad's point about logic is extremely important and for that reason I'll leave it to the end.

I'll turn for the moment to Christopher and John who both think, for different reasons, that I overplay the role of human language in the understanding of the human mind. Christopher thinks I err on one side, because I'm not willing to allow that what goes on in the innards of the computer can count as language; and John thinks that I err on the other side because I leave out of my account of language the personal private mental realm which John thinks is the thing that language exists to communicate. Now, I think that my definition of mind in terms of symbols, in fact, contains all the good points in the other definitions. No doubt, all the other symposiasts feel the same about their definition; but let me explain why I think this about mine.

Wad is right in thinking that there is a very close connection between mentality and having goals, but I think that not any old goal will do as proof that you have a mind. I think only long-term goals of a rather special kind are proof that you have a mind, and goals of that kind can only be formulated in language. Conversely, an activity is only a linguistic activity if it is the activity of a being which is capable of having long-term goals of that kind.

Similarly with regard to plans, a very important part of the

144

notion of mind is being able to plan; but I think only certain kinds of plan are an indication of having a mind, namely symbolic plans and not the simple kinds of plans that dogs and apes can make. Conversely, again, only somebody who is capable of plans of the appropriate kind can be genuinely said to be using symbols.

Now for autonomy. I think that to possess a language or something like it is necessary if one is to be able to have and formulate the types of reason for action which are characteristic of autonomous agents, and I think again that only agents with a certain degree of autonomy can be said to have a language at all. Like Christopher, and unlike John, I don't think that an agent has to be so autonomous as to be indeterministic in order to have a mind or to have a language, but I do think that it has to have a degree of autonomy.

What about consciousness? None of us has chosen it as the defining feature of mind, but obviously it plays a great part in human mentality. I think that the consciousness which is essential if one is to have a mind, is self-consciousness. I don't think that consciousness in the sense of perception gives one a mind because animals perceive, and I don't think animals have minds in the sense that we have them. But human beings are self-conscious and this is part of what is involved in having a mind. Now, I think that self-consciousness is connected with language; indeed, as I have argued earlier, one cannot be self-conscious unless one has a language. This is because there is no way of distinguishing between knowing that something is the case, and knowing that one knows that something is the case, between being in a certain state and knowing that one is in that state, unless one has a symbolic as well as a non-symbolic way of expressing that knowledge and expressing that state. Again, conversely, it seems to me that any being who has a language must be capable of observing rules, and that only beings capable of self-consciousness can do the appropriate sort of self-correction that is involved in really following rules and not just being governed by rules.

I've spoken a lot about language, but I did in fact define mind, at least when I was on my best behaviour, with reference to the ability to operate with symbols rather than with language, and I left it open whether there might be other symbolic outputs which would indicate minds. So I should, I suppose, answer the question: 'What else other than language would I take as being an output which indicated the presence of a mind?' It isn't easy to think of something, but I don't think it's impossible. If we came across a group of organisms which had no language, but which used money as we do, then we should say those organisms had minds. I'm not sure whether or not it is conceivable that there should be a social organisation which had money but did not have language, but if it is conceivable, then I would be prepared to accept it as an indication that the organisms in question had minds. This is because of the abstract nature of money and its use in embodying long-term goals and its usefulness in the preservation and expansion of autonomy, but essentially because it is symbolic and in very much the same way as language is.

I think it's worth developing the deep analogy between money and language — among other things, it brings out some of the limitations in the early work in artificial intelligence in simulation of language. In various airports, you can find machines which change money for you. You put a £5 note in at one end and you get five £1 notes, or 500p out of the other end. Now, the input and output to these machines is genuine money. In the same way, the input and output of Christopher's computer is genuine English, when he plays with it his game 'Waiting for Cuthbert'. But, of course, the money doesn't derive its value from what goes on inside the money-changing machine; it doesn't have any value apart from the resources, the labour and the promises of human beings who use the money. Again, the processes in the machine which mediate between the £5 going in and the £1 notes coming out don't themselves have any value except independently as bits of hardware. Similarly, the English which is typed into Christopher's computer and the English which comes out of it are genuine bits of language, but they don't get their meaning from what goes on inside the computer between the input and the output, they get their meaning from the activities, the knowledge, the experience and the conventions of the human beings who use the language. Again the processes in the computer which mediate between the question: 'Will Cuthbert come?' and the answer: 'No, he never will', these processes don't themselves have meaning.

I think that the analogy is to that extent exact, but of course, I wouldn't want to push it too far. That would be unfair to Christopher, because I think it's absolutely clear that if one wanted to understand the monetary system, one wouldn't learn anything at all by putting together or taking to bits the changing machine, whereas I do believe, as Christopher does, that we can learn a great deal about the nature of language by writing programmes to simulate natural languages.

The analogy also has its uses in answer to John's question. Value isn't conferred on money by any private and spiritual act of the mind, but by a set of public and communal conventions. Similarly with meaning, the attempt to confer meaning by a private introspective act is as futile as the attempt to give value to a currency by a naked act of the will.

I turn to Wad, from language to logic. Logic is in one way a more sublime expression of mind than language; in another way it has nothing special to do with language at all. Let's take a simple logical truth, namely: 'If either p or q, and not p, then q'. That is a logical truth, or a pattern for an indefinite number of logical truths. Now, you can consider that at four levels.

First of all, there is the way in which this law of logic applies to the world: there has never been an event which was a violation of it, and there never could be such an event. To that extent, logic governs the world as well as our minds, though that is perhaps a misleading way of putting it.

Secondly, animals like dogs can apply logic. I'm told that if a police dog is tracking a criminal and comes to a fork in the road, it sniffs at one fork and if it doesn't find the track there, then it immediately goes on to the other road without any further sniffing. If that is true then the dog is, as it were, making an application of the logical law. But I wouldn't want to say that the dog knew any logic, because it doesn't have the ability to express this truth with the generality with which I've expressed it − it can only apply it in particular cases.

Thirdly, normal human beings know simple truths of logic and express this knowledge in their use of words like 'either', 'or', 'not', 'therefore', 'but' over a wide variety of topics.

Finally, there are people like Aristotle and Frege who formalised this knowledge that we all have intuitively and turned it into a branch of mathematics. Now, it's only in sense four that there are these alternative logics that Wad is talking about. One can get various different systematisations of the logical truths of the propositional calculus, but the possibility of giving these various formulations is in itself not much more interesting than the fact that you can write nine as IX or as 9, and the existence of these alternative formulations doesn't mean that there is any possibility of dispensing with logic as it applies to the world any more than the possibility of Roman numerals means that we might count in a different way.

You asked: if a logic excluded a conclusion that you want to get, should you reject it? We have to ask in turn: do you want to get the conclusion any old how, or do you want to get only true conclusions by valid methods? Assuming that it's the latter, and that what you want is a formal system, then you have to choose a logic which is consistent. If it has been proved consistent, you will know that you won't ever be led by the logic from true premises to a false conclusion. If you don't care whether the system is consistent, you might as well go in for wishful thinking. Logic is indeed a tool, but you want a tool that will lead you only to true conclusions from true premises, and to make sure that your logic is of that kind may demand quite hard work.

Questions to Lucas

LONGUET-HIGGINS. In your talk yesterday, you said: 'Any view of the universe large enough to include the fact that minds exist will itself have to be so large, so rational, and so personal, as to deserve the appellation of "God" '. Can I be right in understanding you to assert that a fully mature science would qualify for that title? If not, what are you actually suggesting?

WADDINGTON. The question that I want to put in a way grows out of that. What I took John to mean, was that any view of the universe broad enough to include the existence of mind, could be taken as a description of God. Now the point I want to raise can perhaps be put, in a rather frivolous form, by asking the question: what was God like before the universe included any living beings or any minds? There could not then be any view of the universe broad enough to include minds. Again, at some time in the far future, say in the year A.D. 2,000,000, if evolution continues until then, there will surely be super-minds of some kind, and what will the nature of God be at that time?

But I should really prefer to put the problem in another way. We have been agreeing that the world has a structure, and that this is a structure in which processes are related by instructions or algorithms. Now, such a system is essentially open-ended. It is one in which things can be created. In fact, one might say that during the course of evolutionary history there have been creative changes in the world, which have brought into being such things as minds, as those are defined by Tony and John. This is another way of making the point which I referred to as a change in the nature of God in my first few sentences. Now John has been laying great stress on autonomy and freedom. I think really the question I want to ask is whether by freedom he means simply complete freedom to do whatever you wish, regardless of circumstances; or whether he would relate his idea of freedom in any way to the creative, open-ended character of the universe. It seems to me that there ought to be some connection between what John refers to as freedom, and the open-ended, creative character of the universe which you might call 'the evolution of God', but I don't know quite what this relation is, and I should like to hear John talk about it.

KENNY. My question is very close to the other two. John, you insist frequently on the rationality of the universe, but there are two different things this might mean and I'm not sure which of them you have in mind. You may mean that the universe is rational in the sense that it is intelligible to rational creatures like ourselves, in which case I think, none of us would be likely to deny this. Or you may mean that the universe is rational in the sense that it exists and develops the way it does for a reason: that is, that there is a final cause for the universe, there is an end or purpose to the universe. Now, there is a big difference between these two senses of rationality and I'd like to know in which sense you think the universe is rational.

The diagram drawn by Dr Kenny (see p. 137) showed only four of the characteristic features of minds. But there are others. In particular there are various forms of creativity, spontaneity and originality, which I should place at a higher level than autonomy, and near the apex. Wad is quite right to think that these are very important marks of mind, which we have not sufficiently considered. Without them, my emphasis on autonomy would collapse into soggy libertarianism. And I certainly should want to disown that. The reason why I've been stressing autonomy, which carries with it always the possibility of being wrong, is because I think the possibility of things going wrong at the level of autonomy is a necessary condition of being right at the level of creativity and of being right in an original non-algorithmic way. That is to say, I do argue against Tony that the liberty of spontaneity implies, in some sense, the liberty of indifference. This is why I am an indeterminist. But although this means that I am defending my colleagues' right to be wrong, the underlying reason why I am so anxious to secure to them the right to be wrong, is that this is a necessary condition of their ever being right. So long as they are being wrong, while I defend their right to be wrong, I am sorry that they are wrong, and try and persuade them of the true opinion; and these amount to more original and creative ideas nearer the apex.

Now, let me try and answer Christopher and Tony together. Christopher is right in having caught me in soft and woolly thinking, as I now look at that passage in my text he pointed out. The antecedent of 'itself' is unclear; do I mean 'God', or do I mean 'view of the universe', and this makes me open to Wad's other question: 'What was God like before the universe included any living beings or any minds?' There is a difficulty here because it is very easy to slip between *ways* of talking – various descriptions – and *what* one is describing. This underlies also the difficulty which Tony is bringing up about rationality. Ideally, I should like to be able to put an absolutely sharp distinction between something there which is referrable to and something here which is my means of referring to it. But when we get into the realms of philosophy and metaphysics, it is very difficult to maintain this distinction. Often what we are talking about is almost constituted, almost but not quite, by what we say about it, just as it is in perception. Often, we are only able to see the things that we were expecting to see. It is quite difficult to drive a wedge in there, just as it is quite difficult for me to draw the distinction, which Tony rightly thought needed to be drawn, between saying that the universe is rational in the sense that it is intelligible to the human reason, and saying that the universe is rational in the sense that it exists for reason. Once this distinction is drawn, then I will accept it and say we first of all come to the conclusion that the universe is intelligible by us; and then when we face some difficult question as when a mother is bereaved and loses her infant, we say: 'Why is the universe like this?' And if we attempt to make out that it was the manifestation of a creative mind, we can ask further questions there. This is an instance where the inference can be made explicit, so that Tony can query it. But it is difficult often to make this sufficiently explicit; just as when we are talking about other people, it is difficult to distinguish the overt behaviour patterns from the mind behind them. Rather we talk of reading a man's mind. If one pushes this distinction through too far, one often lands himself in an entirely unnecessary problem of other minds. Similarly, when we are talking about God, my confusion between talking about a world view and talking about the God in terms of whom I understand the world, is not quite a legitimate, but often almost a necessary, conflation of terms.

Finally, Christopher asks me about a world view sufficiently large to include us being necessarily, therefore, one which is based on the existence of God. I would like to be able to produce a straightforward argument of that sort — or perhaps even quote a scholastic tag about explanations necessarily being as strong and as powerful as all things to be explained. But I can't produce a positive argument — all I can do is hand-wave in this direction, and then hope that whenever an alternative hypothesis is put forward I shall be able to refute it. I tried to show this scheme of refutation originally in that Gödelian argument, and then yesterday when I was saying that Monod's account, just because it doesn't embrace enough, can be shown in some sense as self-defeating and incoherent; but whether I can do this always is a matter on which you may have legitimate doubts.

Let me now change my tone of voice. As I was to be the last speaker my colleagues didn't want to give me the last word as against them, but for them. And now I'm no longer quarrelling with them but speaking for us all, saying a certain number of 'thank-yous'. Some 'thank-yous' are really to the Gifford Committee rather than to you here, for bringing the two of us, the Southcountrymen, up to Edinburgh for many repeated, valuable and enjoyable visits; and from all of us for bringing us together, in creating new friendships and giving all of us a great deal to ponder and to think about. But to you here, we want to give another pair of thanks. We were very alarmed when we got the Principal's letter — we shy, retiring dons were being asked to come out of our ivory towers and to go into 'show-biz'. We had got to be entertaining and we had got to communicate — that, I think, is the trendy word. As Professor Longuet-Higgins was saying a couple of days ago, communication is something which is essentially two-way, and something which comes to dons very difficultly — we normally sit in our studies and we write down, we cross a word out there, and there is no feed-back — this is often just as well: it is a one-way process, and that is very easy at the giving end; but here it has been very much a two-way process, and we have depended greatly on your responses. We have found it very valuable to be, as it were, carried along by you; sometimes you even laughed at our jokes; we think that sometimes you followed our arguments; and we have been very largely carried along by the mere fact of your being here. When we were planning these lectures, with dire warnings from our colleagues and from the Gifford Committee and from previous lecturers, we were told that the curve of attendance was a negative sigmoid one, and that after the second lecture we were going to be talking to five dear old ladies knitting. Well, I can see that some of you are ladies and I can see that some of you are old; whether any of you are dear or not, is something which is not open to immediate inspection, though I am sure that you are. I'm reasonably confident, from looking at you many times, that none of you have been knitting. And we are all absolutely certain that there hasn't been only five of you. And for this we thank you.

Index of Names

Aristotle, 23, 25, 27, 28, 51, 52, 53, 147
Auerbach, Dr Charlotte, 12
Austin, John, 19
Bennett, J., 93
Bertallanfy, 75
Black, Stephen, 73
Bohm, David, 41
Bower, Tom, 1
Bruner, Jerome, 32, 36–7, 41
de Chardin, Teilhard, 108
Chomsky, Noam, 1, 4, 35–6, 38–9, 40, 41–2, 45, 49, 67–70, 92–3, 97, 98–9
Le Gros Clark, W., 71
Coleridge, 55
Darwin, C., 21, 27, 74, 108
Descartes, 48, 50, 51, 54, 60
Eddington, Sir Arthur, 110
Einstein, A., 40
Frege, Gottleb, 48–9, 147
Freud, S., 72–3, 85, 88
Godel, 53, 54, 58, 61, 139
Hare, R. M., 25–6, 57
Hardy, Sir Allister, 20–1
Hill, Christopher, 108
Hockett, 99
Hume, David, 17–18, 50, 87
Huxley, Julian, 84
Kant, 9, 18, 25–6, 28–9, 57
Kauffman, Stuart, 81, 83, 137–8
Keenan, Dr Edward, 68
Koestler, Arthur, 2, 71, 72–3, 85

Lamarck, 21
Leakey, Robert, 6
Lenneberg, 91–2, 97–8
Locke, John, 9
Lyons, J., 91–2
Mehler, Jacques, 5
Monod, Jacques, 123–4, 127–33, 135, 150
Needham, 75
Neumann, J. von, 15
Newman, Stuart, 81, 83
Newton, 25
Papert, Seymour, 3–4
Piaget, Jean, 1, 3–4, 32, 37–8
Plato, 48–9, 51, 52, 55, 57, 62
Popper, Sir Karl, 76, 79, 98
Premack, D., 6
Pythagoras, 57, 59
Russell, Bertrand, 28, 138
Schroedinger, 78
Searle, J., 93
Shannon, C. E., 77, 89
Skinner, B. F., 31, 39
Maynard Smith, John, 96, 102
Tarski, 139
Thorpe, W. H., 70
Weaver, Warren, 77
Whitehead, A. N., 41, 75
Whorf, 40
Wittgenstein, L. von, 93, 104
Woodger, 75